Combustion of Sprays of Liquid Fuels

PROFESSOR ALAN WILLIAMS,
B.Sc., Ph.D., C.Eng., M.Inst. F., F.R.I.C., F.Inst. Gas E.,
Department of Fuel and Combustion Science, The University of Leeds

Elek Science   London

© Paul Elek (Scientific Books) Ltd., 1976
First Published in Great Britain in 1976 by
Paul Elek (Scientific Books) Ltd.,
54-58 Caledonian Road,
London N1 9RN

ISBN 0 236 31044 5

Printed in Great Britain by Unwin Brothers Limited
The Gresham Press, Old Woking, Surrey, England.
A member of the Staples Printing Group.

Combustion of Sprays of Liquid Fuels

Contents

Preface

The development of spray combustion has effectively followed two separate paths, one concerned with engine applications, in which the aerospace applications have dominated the field; the second has been concerned with stationary equipment such as furnaces and boilers. Generally, textbooks have been concerned with one area or the other and the object of this book is to outline the fundamentals of the combustion of sprays in a unified way which may be applied to any existing or future technological application.   During the last decade the necessity of controlling the emission of pollutants has assumed greater significance in all aspects of combustion.   More recently the requirements have been for pollution control and increased combustion efficiency.   Doubtless, in the not too distant future, it will be necessary to burn fuels which are of low or variable quality or synthetic fuels having properties differing greatly from present-day fuels.   For all these developments a greater understanding of spray combustion is required.

This book has been written at a level suitable for students at undergraduate or postgraduate level undertaking courses in fuel, combustion or energy studies.   The emphasis has been towards the fundamental aspects although some present day applications have been outlined.

I am indebted to a number of individuals and firms for assistance and the provision of material, namely Dr D. Anson, Dr H.A. Cheetham, Dr B.D. Edwards, Mr W. Horne, Dr P.J. Street, and Dr J. Swithenbank and CEGB, Fuel Furnaces Ltd., Hamworthy Engineering Ltd. Combustion Division, Urquhart Engineering Co Ltd. and Weishaupt (UK) Ltd.

I am also indebted to a number of people for stimulating discussions concerning spray combustion and, in particular, Dr D. Bradley, Dr N. Chigier, Dr N. Dombrowski and the members of the department of Fuel and Combustion Science. I would also like to thank Miss J. Hockin for typing the manuscript and my wife for her patience and understanding.

1 Spray Combustion as a Source of Energy

1.1 THE GENERAL NATURE OF SPRAY COMBUSTION

The combustion of sprays of liquid fuels is of considerable technological importance to a diversity of applications ranging from steam raising, furnaces, space heating, diesel engines to space rockets. Because of the importance of these applications spray combustion is responsible for a considerable proportion of the total energy requirements of the world, about 30% in 1974.

Spray combustion was first used in the 1880s as a powerful method of burning relatively involatile liquid fuels. The basic process involved is the disintegration or atomisation of the liquid fuel to produce small droplets in order to increase the surface area so that the rates of heat and mass transfer during combustion are greatly enhanced. Thus the atomisation of 1 cm^3 of liquid into droplets of 100 μm diameter increases the surface area by a factor of 1228.

A burning spray differs from a premixed, combustible gaseous system in that it is not uniform in composition. The fuel is present in the form of discrete liquid droplets which may have a range of sizes and they may move in different directions with different velocities to that of the main stream of gas. This lack of uniformity in the unburnt mixture results in irregularities in the propagation of the flame through the spray and thus the combustion zone is geometrically poorly defined.

Flames used in industrial applications are highly complex systems because of various complicating factors such as the complex flow and mixing pattern in the combustion chamber, the heat absorption during the combustion process, and the non-uniform size of the particles. For purposes of analysis the assumption is often made that the system is uni-dimensional. Thus the flame can be considered as a flowing reaction system in which the properties of flow, temperature, etc. vary only in the direction of flow and are constant in any cross section perpendicular to the direction of flow.

The general nature of the processes involved in spray combustion in such an idealised case for the combustion of a dilute spray is shown in Figure 1.1.

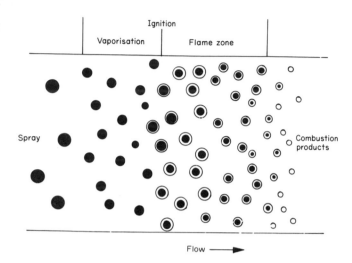

Figure 1.1.   *A diagrammatic model of idealised spray combustion*

In this case the individual droplets that make up the spray burn in a surrounding oxidising atmosphere which is most commonly air.   This is also clearly shown in Plate 1.1 which shows a simplified (flat) spray system.   This plate also demonstrates the other major features of a spray flame, namely atomisation, air entrainment and flame stabilisation. The flame front is visible as a diffuse zone across the upper part of the plate.

It is clear that for any detailed understanding of the process of spray combustion it is necessary to have an adequate knowledge of the combustion of the individual droplets that make up the spray, since a burning spray may be regarded as an ensemble of individual burning or evaporating particles.   However, it is also necessary to have a statistical description of the droplets that make up the spray with regard to droplet size and distribution in space.

2

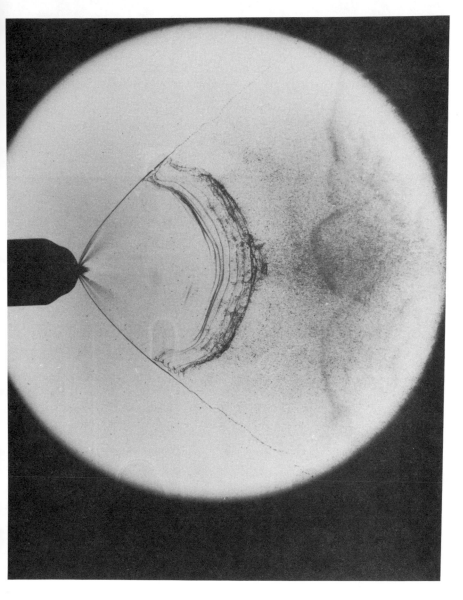

*Plate 1.1.   The essential features of a burning spray.   This particular
spray flame is produced by means of a fan atomiser.*

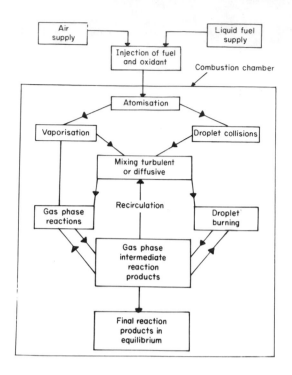

Figure 1.2. *The major combustion processes in a combustion chamber.*

The essential stages involved in spray combustion are outlined in *Figure* 1.2. The fuel is transmitted from the fuel storage tank (by a fuel handling system that incorporates pumps, filters and some means of control) to an atomiser by means of which the fuel is atomised into small droplets; these droplets are usually injected directly into the combustion chamber where they burn. The combustion process is very complicated because, to a large extent, the mixing of the fuel and oxidant takes place inside the chamber and thus the mechanics of the mixing process play an important role. This mixing process is controlled to a large extent by the geometry of the combustion chamber, the spatial distribution and momentum of the injected spray, the direction and momentum of the air flow and the influence of any flame stabilisation devices. Consequently, the atomiser and combustion chamber should be designed as an integrated unit rather than as independent items.

4

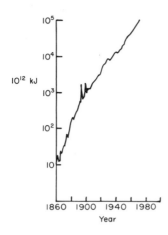

10^{12} kJ

Year

Figure 1.3.   *Growth in world oil consumption expressed*
in terms of energy.   Some energy
interconversion factors are given
in Appendix 1.

1.2   THE SOURCES OF LIQUID FUELS USED IN SPRAY COMBUSTION

The liquid fuels used at the present time are almost exclusively
the fuel oils.   The term 'fuel oil' means different things in
different countries but essentially it covers the range of products
from gas oil to extremely viscous products of high molecular weight.
This group encompasses both diesel fuels and industrial fuels used in
boilers and furnaces.   Their major source is crude oil, but liquid
fuels can also be produced from coal as well as oil shale and tar
sands.

In view of the increase in world petroleum use, as shown in
Figure 1.3, and the finite nature of oil reserves as illustrated, in
Figure 1.4, there has been considerable interest in the alternative
sources, although, for the time being, crude oil must be the dominant
source.   Table 1.1 lists the present day fossil fuel reserves.   The
general indications at the present time are that the world total of
ultimately recoverable oil is of the order of 300 x 10^9 tonnes.   This
includes 40 x 10^9 tonnes already recovered and some 90 x 10^9 tonnes in
proved reserves;   the rest remains to be discovered.

5

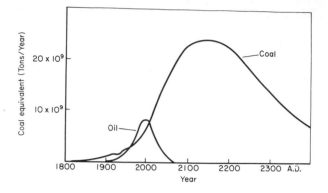

Figure 1.4.   *Comparison of possible future usage of oil and coal. The area under each curve represents their respective world reserves.*

On this basis, Table 1.2 indicates the possible lifetime of crude oil based on (a) the present day demand, and (b) a continuing expansion of oil demand, which has increased at the rate of 5 to 10% per annum. Whilst the future growth rates remain uncertain it is clear that by the end of the present century alternative sources of liquid fuels must be developed.

Table 1.1.   Estimates of world fossil fuel reserves

(10^9 tonnes oil or oil equivalent) *(Energy for the Future, 1973)*

Oil	80	-	300
Shale/tar sand	97	-	200
Coal	130	-	4800

Table 1.2.   Estimated lifetimes of possible oil supplies (years)

	Life at 1974 consumption rates	Life at future consumption rates
Oil	30 - 100	16 - 35
Shale/tar sand	40 - 200	20 - 60
Coal	50 - 2000	30 - 250

1.2.1 Fuel Oils from Crude Oil

The main source of liquid fuels at the present time is crude petroleum which occurs naturally in the earth's strata. Crude oil consists essentially of hydrocarbons together with smaller quantities of sulphur, oxygen and nitrogen containing hydrocarbons and some organo-metallic compounds, particularly of vanadium. Gaseous, liquid and solid (or semi-solid) compounds may be present in crude oil and these are separated at the well-head and during the refining processes to give a range of liquid products, some of which are used as fuels for spray combustion. The properties of the liquid fuels produced are markedly dependent upon the source of the fuel, the nature of the refining operations and the method of blending used to produce the final product.

The hydrocarbons present in crude oil have differing boiling points and are separated by the process of distillation into a range of primary products. The nature of many of the final products, particularly the fuel oils, is determined by the chemical composition of the crude oil that is distilled. Because of the multiplicity of the molecular species present in oil, crude oils (as well as the products) may be classified in terms of the concentrations of broad chemical groupings, namely paraffinic, naphthenic, aromatic or asphaltic. During the process of distillation and other refining operations these products are distributed amongst the final products according to their properties; so also are the sulphur-containing hydrocarbons, etc. and inorganic components.

A typical but simplified refinery flowsheet is illustrated in Figure 1.5 although it must be recognised that very many variants are possible depending upon the type of crude oil and the range of products required. In particular the pattern of fuel usage in the area served by the refinery plays a key role in controlling the exact processes used. In addition the refinery scheme is also dependent upon its involvement with demands for petrochemical feedstocks.

The raw crude oil is fractionated in the crude distillation

7

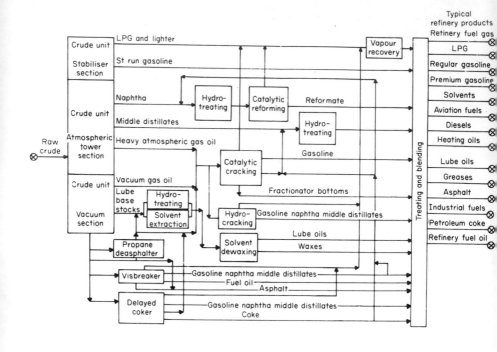

Figure 1.5. *A typical refinery flowsheet.*

unit into distillate and residuum streams. Typically, in a two-stage
unit, an atmospheric tower produces middle distillates and lighter
fractions, while a vacuum section produces a heavy gas oil cracking
stock and other streams for lubricating oil production, etc. Before
processing, the salt and water concentrations are reduced and then the
crude is flash distilled, producing gases, gasoline and a number of
sidestreams. These sidestreams, after additional purification stages,
yield aviation kerosines, diesel fuels and gas oils. The major
finishing process used for these products is hydrotreating. In this
process mild hydrogenation conditions are employed to reduce the
concentrations of sulphur, oxygen and nitrogen compounds as well as that
of the unsaturated compounds. A number of properties are improved by
this process, particularly odour and stability. The stripped
atmospheric bottoms are reheated and passed to the vacuum tower. Here
more (vacuum) gas oils are produced. The vacuum residuum can be used
directly or further processed. In the latter case the residuum may be
processed in a visbreaker to produce essentially a gas oil product

directly available for sale. Alternatively, it may be charged to a delayed coker to produce gas oil and lighter fractions as well as coke. Thus, apart from these two processes, the distillation process provides the whole range of fuel oils, since the final products are blended from straight distillates, residues and by-products from other refinery operations. A number of additives may be incorporated into the final product; in the case of fuels used for spray combustion the additives normally present are anti-oxidants, whilst in the case of engine fuels, pour point depressants, antiwear additives and antirust additives may be present. Other additives to reduce pollutant formation may be added prior to use; these are discussed in Chapter 7. Further details of these processes are outlined in standard textbooks on refining (e.g. Hobson and Pohl, 1973).

1.2.2   Oil from Shale and Oil Sands

Extensive deposits of oil shales and oil sands (tar sands) exist throughout the world. The reserves of oil from such sources are considerable, as indicated in Table 1.1. Of major significance are the oil sands in Canada (Alberta), USA and Venezuela and the oil shale deposits in the USA and Brazil. Oil sands consist of deposits containing heavy hydrocarbons which are the same as those in conventional oil but have high viscosities, i.e. they are tar-like. Oil shales are significantly different, since they contain no free oil, but consist of a solid mixture of organic compounds, called 'kerogen', which decomposes on heating, yielding a light shale oil.

Production of oil from shale or oil sand involves extracting the raw material, processing it to extract the hydrocarbons and then converting the crude oil so produced into a form in which it can be used.

The methods used depend upon the nature of the deposit. In the case of oil shale, extraction involves a mining operation and the product is then heated to about 400°C. The liquid product obtained has many of the properties of a conventional crude oil, except that it has a high viscosity and a high nitrogen content. An alternative method involves the *in situ* conversion of the oil shale to oil ; this would obviate the mining operation , but so far these techniques have been unsuccessful.

In the case of oil sands, two extraction methods are generally applicable, depending upon the nature of the deposit. Firstly, hot water or steam may be introduced into the formation to reduce the viscosity of the oil so that it can be pumped out in the conventional way. This is particularly suitable for the deeper oil sand deposits in which the deposits have fissures or are sufficiently permeable for the

oil to flow to the production well.   The second technique is similar to
the oil shale operation in which the oil sand is recovered by a mining
operation and then processed using hot water or steam and diluents.
Even so, the recovered oil has a low specific gravity, high viscosity
and high sulphur content necessitating upgrading before use.

1.2.3   Liquid Fuels from Coal

The production of oil from coal has been the subject of
considerable research effort, particularly during the Second World War,
and is receiving considerable interest at the present time.   There are
three basic processes available:

(a) *Methods based on the Fischer-Tropsch process*

Here the coal is gasified by conventional processes (these are
outlined in Hottel and Howard, 1971) to a mixture of carbon monoxide and
hydrogen.   These are reacted catalytically in the Fischer-Tropsch
process producing hydrocarbons thus:

$$n \text{ CO} + (2n+1)H_2 = C_n H_{2n+2} + nH_2O$$
$$2n \text{ CO} + n H_2 = (CH_2)_n + n CO_2$$

together with alcohols

$$n \text{ CO} + 2n H_2 = C_n H_{2n+1}OH + (n-1)H_2O$$

Such a coal into oil process is currently operated by the South African
Coal, Oil and Gas Corporation (SASOL) in which the CO/H_2 mixture is
reacted in two ways, in the Lurgi (Arge) process and the Synthol process.
In the former a pelletised iron fixed-bed catalyst is used and a wide
range of hydrocarbons is produced including gasoline and furnace oil (a
gas oil equivalent) together with a range of waxes.   In the Synthol
process, a fluidised iron catalyst is used and gasoline, gas oil and
alcohols as well as other products are produced.   The Synthol process
operates at higher temperatures, resulting in comparatively low yields
of heavier oils and waxes.   The process is very flexible, and by
changing the catalyst composition the product spectrum can be varied to
suit the demand of the end use.   In addition, if the sole objective
were to produce liquid fuels for spray combustion the liquid hydrocarbons
and the alcohols do not need to be separated before use.

(b) *Coal Pyrolysis*

During the carbonisation processes in the manufacture of coke and
the older coal gas processes the products include coal tar.   Some of

this is used directly as a fuel and the rest is distilled to give a series of fuel oils, the coal tar fuels and also a benzole fraction which is blended with gasoline. The coal tar fuels are designated CTF 50, 100, 200, 250, 300 and 400, the number being its recommended atomisation temperature in degrees Fahrenheit. These fuels can be burned as spray flames in the usual way although they are highly aromatic fuels and produce highly luminous flames. Some of the important properties of some coal tar fuels are listed in Table 1.3.

Table 1.3. Properties of some coal tar fuels

Type	CTF 50	CTF 100	CTF 200	CTF 250
Viscosity, Red I. $37.8°C$ ($100°F$)	60	100	1 200	4 000
Gross calorific value, kJ kg^{-1}	39 600	39 000	38 300	38 000
Flash point, $°C$	82	93	99	103
Sulphur %	0.75	0.8	0.9	1.0
Specific gravity	1	1	1.1	1.2

The yield of liquid fuels by this route is small (about 8% wt) but can be increased to some 75% by the use of hydrogenation. The process relies on the rapid heating of finely divided coal to drive off the volatile components, both liquid and gaseous. Often fluidised bed reactors are employed and staged so that the pyrolysis conditions become increasingly more severe with each stage. The best known method here is the COED process where a low sulphur fuel oil is produced by hydrotreating the liquid product. This, and a number of other similar processes, have been discussed by Hottel and Howard (1971). The major disadvantage of coal pyrolysis routes to syncrude is that a considerable amount of char results, some 50-60% wt. Thus, the processes must be used in conjunction with power generation or gasification. A number of coal-based complexes have been proposed, e.g. the NCB coalplex.

(c) *Solvent Extraction*

Solvent refining of coal can be undertaken by adding hydrogen directly to the coal. The coal is slurried and dissolved in a solvent under high temperatures and pressures and hydrogenated. The ash is removed by filtration and the liquid fuel produced can be burned as such. Alternatively, it can be further hydrogenated to yield synthetic crude for further refining.

1.3 THE NATURE AND PROPERTIES OF FUEL OILS

The chemical and physical properties of the fuel oils largely determine their particular application.   Thus, automotive and small domestic units require an oil that is easily handled and atomised. On the other hand, large plant can handle highly viscous oils that may require preheating.

The major properties relating to the handling, atomisation and combustion of liquid fuel sprays are briefly outlined below.   The fuel properties may be determined by standard tests which are outlined in the British Standard Specifications, The Institute of Petroleum (IP) publication, *Standards for Petroleum and its Products* and the ASTM *Manual on Measurement and Sampling*.

Viscosity

Viscosity is a measure of resistance of a fluid to flow and therefore affects the energy required to pump a fuel through a pipe, and so it markedly affects the process of atomisation.

Viscosity is a function of temperature.   This is illustrated in Figure 1.6 which shows the variation of the viscosities of some typical

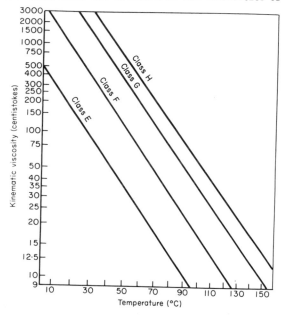

Figure 1.6.   *The variation of viscosity with temperature of some typical commercial fuel oils.   The letters refer to the British Standard designations as given in Table* 1.5.

industrial fuels with temperature.

A number of arbitrary methods, e.g. Redwood, Saybolt methods, have been used to measure viscosities but the kinematic method, which measures the flow rate in a standard U-tube viscometer, is now being widely used. The values are expressed in centistokes and in the U.K. are given at 82.2°C (180°F). The inter-relationship between kinematic viscosity and other common scales is indicated in Appendix 1.

Pour Point and Cloud Point

The cloud point is the temperature at which a haze appears when an oil is cooled and it indicates the onset of the formation of wax crystals which can block filters. The pour point is several degrees lower than the cloud point and is defined as the temperature which is 2.8°C (5°F) above the temperature at which the oil just fails to flow when cooled. This is due in general to the separation of wax from the oil.

Sulphur Content

Sulphur, in the form of organo-sulphur compounds, exists to some extent in all liquid fuels. This influences to some extent the calorific value but the major significance of the sulphur content is on the formation of sulphur oxides and is discussed in Chapter 7.

Carbon Residue

This gives a measure of the quantity of solid deposits obtained when medium or heavy fuel oils are heated so that evaporation and pyrolysis takes place. Most of the lighter components evaporate but the higher molecular weight compounds decompose to yield a carbonaceous deposit. The Ramsbottom test is commonly used for fuel oils.

Ash

For most petroleum fuels the distillate fuels contain a negligible amount of ash, but the residual fuels contain a very small amount, up to about 0.1%, depending on the grade.

Water and Sediment

With distillate fuels, the amount of water and solid sedimentary material is negligible. The heavier fuel oils can retain small amounts of these materials in suspension, but the amount of water is unlikely to exceed 0.5% or sediment 0.05% even in the most viscous fuel.

Specific Gravity

This determines the mass per unit volume and is thus important in the metering of fuels and the energy liberated per unit volume of fuel burned.

A number of other properties are of major importance but are not relevant to spray combustion.

1.3.1   The Classification of Fuel Oils for Diesels and Turbines

Gas Turbine Fuels

Gas turbine fuels used for aviation purposes are essentially straight run products obtained from the distillation of crude oil.   For aviation purposes, essentially two grades are available, a kerosine type fuel (AVTAG) and a wide cut gasoline type (AVTUR) fuel.   Industrial and marine gas turbines are usually designed to burn on distillate fuels ranging from naphtha to industrial gas oil, whilst a limited number of gas turbines operate on selected residual fuel oil.

The specifications for aviation fuels are very stringent and are covered in the U.K. by the Ministry of Defence (D Eng RD) specifications and in the U.S. by the Department of Defense (US-MIL), although similar specifications are quoted particularly by ASTM and other bodies.

A brief summary of the major properties of some typical fuels are given in Table 1.4.

Diesel Fuels

Diesel fuels are basically straight run products obtained directly from the distillation of crude oil.   In addition, as indicated in Figure 1.5, they contain varying amounts of cracked distillates so as to increase the yield.

The distillate fractions, the gas oils, vary widely in composition but usually the aromatic contents lie in the range 15-30% by weight; the paraffins may vary from 25-70% by weight whilst the naphthenes may vary from 4-60% depending upon the source.   Obviously these variations have a marked effect on fuel properties, particularly the cetane number. In addition, the fuel oil may contain residual fuel oils which are complex mixtures of higher hydrocarbons which may be paraffinic or asphaltic in nature.

A number of classes of diesel fuels are available, these broadly are:

Table 1.4. Properties of some engine fuels

Requirements for diesel engine fuels. British Standard 2869

Fuel type	A1 for high speed automotive engines	A2 for low speed engines	B1 for Marine use	B2 for Marine use
Viscosity at 37.8°C, cS	1.6 - 6.0	1.6 - 6.0	-	-
Cetane no., min.	50	45	35	-
Carbon residue, max., % wt.	-	-	0.2	1.5
Flash point (Pensky Martens closed cup), °C, min.	55	55	66	66
Ash content	0.01	0.01	0.01	0.02
Sulphur content, % wt.	0.5	1.0	1.5	1.8

Requirements for aviation turbine fuels

Fuel type	AVTAG wide cut	AVTUR kerosine	AVCAT high flash
Viscosity at -34.4°C, cS		15	
Smoke point, mm		20	20
Flash point, °C, min.		38	60
Total sulphur, % wt. max.	0.4	0.2	0.4

(a) A volatile distillate fuel oil mainly used for automotive (road transport) diesels.   These have a boiling range of 200-350°C.
(b) A distillate fuel oil of lower volatility for engines in industrial or heavy mobile diesels (tractors, earth moving equipment, railway locomotives); it can also be used for industrial gas turbines.   It has a boiling range of c. 200-370°C.
(c) A distillate fuel similar to (b) for marine applications.
(d) Fuels containing a certain amount of residuum for larger engines of low or medium speeds with applications in marine or industrial generation fields.   It also finds some application in industrial gas turbines.

The specifications of diesel fuels vary from country to country but the two most widely adopted are the British Standard and the ASTM standard.   The British Standard specifications are outlined in Table 1.4; details of other specifications are summarised by Allinson (1973).

1.3.2   The Classification of Industrial Fuel Oils

The fuels considered here are entirely those used in spray combustion application and range from distillate fuels of the gas oil type used in domestic heating through to heavy residual fuels.   The descriptive term 'domestic heating oil' is usually applied to a gas oil type product, having a boiling point range of 160-370°C, used with spray combustion type domestic heating equipment.

The residual fuel oils are basically the residue resulting from the removal of the more volatile constituents during distillation but, in addition, a variety of other products are blended with it.   The properties of the residual fuels so produced thus depend largely upon the source of the crude oil and the composition of the fuel oil blends. From the spray combustion viewpoint the key properties are the viscosity, the volatility and the ash and sulphur contents.   Other important factors relating to the blockage of fuel filters, erosion of burner tips and pumps, and pumpability are the water content, inorganic material such as rust, the acidity and stability to sludge formation and the cloud point.   Fuel oils are covered by specifications determining the upper limits of the key properties, the more important specifications being the British Standard Specification 2869:1970 and the ASTM D396-69. Some typical properties of industrial fuels in the U.K. are given in Table 1.5.

Table 1.5. Properties of some industrial fuel oils
used for spray combustion

British Standard Class [*]	D	E	F	G	H
Fuel type	Gas oil	Light fuel oil	Medium fuel oil	Heavy fuel oil	
Viscosity at 37.8°C,St	1.6-3.0				
at 82.2°C,St		12.5(max)	30(max)	70(max)	115(max)
Typical gross calorific value, kJ kg^{-1}	45 600	43 500	4 300	42 600	42 400
Flash Point (Pensky-Martens, closed) min.,°C	55	66	66	66	66
Sulphur, % wt. max. Typical S, % wt.	1.0	3.5	4.0	4.5	5.0
Ash, % wt. max.	0.01	0.1	0.15	0.2	0.2
Specific gravity (typical values)	0.84	0.93	0.95	0.97	0.97

[*] British Standard 2869

2 The Properties of Sprays

The analysis of sprays of liquid fuels produced by oil burner atomisers in terms of droplet size distribution, spray angle and spray pattern is important in all applications of fuel spray combustion. In general, only a limited number of atomisers produce reasonably mono-sized droplets, and atomisers in practical combustion chambers produce directly or indirectly a spray having a spectrum of droplet sizes.

2.1 EXPERIMENTAL METHODS FOR DETERMINING DROPLET SIZE DISTRIBUTION

From the practical point of view it is important to be able to obtain an experimental determination of the droplet sizes in terms of numbers of droplets in a particular size range. However, it should be noted that the size distribution of a spray will vary with distance from the atomiser, because the acceleration (or deceleration) of each droplet is an inverse function of droplet diameter. Therefore, the spectrum of a polydisperse spray will vary with distance from the atomiser. In addition droplet-droplet collisions may occur, especially in dense sprays under turbulent combustion conditions.

The most commonly used sizing techniques are as follows:

The Frozen Drop and Wax Method

In the 'frozen drop' technique the atomised spray is directly injected or passed into an alcohol bath which is maintained at the temperature of dry ice; alternatively they are injected directly into liquid nitrogen. The frozen particles so produced can then be sized either simply by sieving, or by photographing the particles and counting them later, using enlarged photographs or by means of an automatic counting technique. Of the automatic counting techniques available, one convenient method involves the use of an image analyser (e.g. the Quantimet Image Analyser) in which the sample may be examined directly by a microscope system or photographs may be sized by an epidiascope arrangement. The image produced by the television camera scanner produces an image of the type shown in Plate 2.1; these signals can be analysed by a central processor to give the size distribution of the particles being studied.

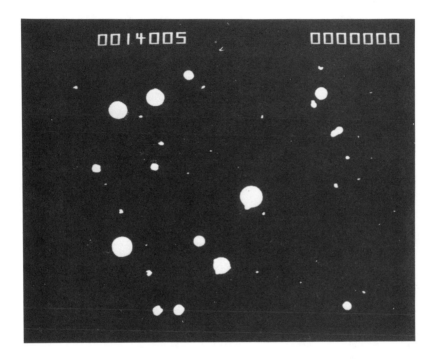

Plate 2.1. Use of the Quantimet to size wax particles produced from an atomiser.

The 'wax method' is dependent upon the fact that paraffin wax, when heated to an appropriate temperature above its melting point, can be used to simulate liquid fuels since it has properties such as viscosity, suface tension, etc. which are similar to many liquid fuels, in particular the aviation turbo-fuels. In the application of this method molten paraffin wax at the appropriate temperature is pumped to the spray nozzle and the wax spray so produced is directed into water in order to solidify the droplets produced. The solid wax particles are collected and sized by passing them through a series of graded gauge sieves.

Microscopic Examination of Collected Droplets

One of the simplest techniques, and for that reason one of the most widely used, involves collecting the sample of a spray on a glass microscope slide and making a microscopic examination of some 500 to 1000 droplets. There are many variants of this technique which are described by Putnam and Thomas (1957). One convenient form involves using a microscope slide coated with a film of magnesium oxide on to which the droplets may impinge. These cause indentations in the MgO film as indicated in Plate 2.2. The sizes of the indentations may be measured as previously indicated but they have to be corrected for the effect of flattening on impaction, the factor of 0.74 being commonly employed. Care must be taken that not more than 1% of the slide is covered with droplets so that the chances of multiple impactions are reduced; a photographic shutter device is convenient here.

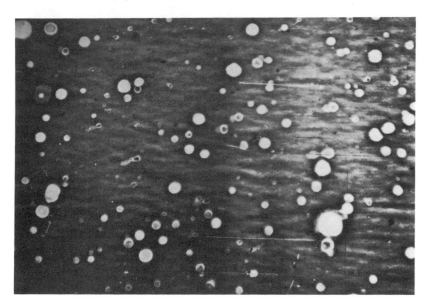

Plate 2.2.   *A magnesium oxide coated slide used for droplet measurements (droplet sizes in the region of 100 μm).*

20

Direct Photography

A further direct method of sizing droplets, but this time *in situ* in the spray, is by direct photography as illustrated in Figure 2.1.

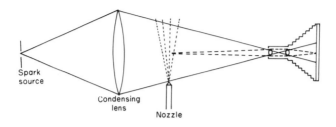

Figure 2.1. *Typical arrangement for photography of a spray.*

This technique is an extremely valuable and convenient method of recording the size distribution of droplets in a liquid spray since the collection and need for introducing an obstruction in the path of the spray are avoided. Essentially the technique involves a short duration light source, less than 1 μs, and a conventional camera arrangement as shown in Figure 2.1. However, great care must be taken in applying this technique to avoid distortion of the size distribution obtained. A development of this technique is laser holography but this involves a lengthy analytical procedure.

Rapid Sizing Techniques

The techniques described so far are laborious in execution and involve lengthy analytical procedure. A number of methods have been devised so as to obtain a rapid indication of the droplet sizes.

One method of directly measuring the sizes and distribution of droplets in a spray is by means of direct electronic counting using a probe inserted into the spray. Such methods have the additional advantage of providing information on the spatial distribution of droplets. A typical example of this technique is the pulse counting technique (Pye, 1970) which measures the droplet size spectrum by depending upon the individual droplets completing an electrical circuit as they pass between two needles positioned an appropriate distance apart. In such electronic counting techniques it is thus necessary to use a conducting liquid as the atomised fluid. By varying the spacing of the needles over the size range of interest and counting the number of droplets at each separation the size spectrum may be obtained.

Usually, an aqueous solution of a salt is used and care obviously has to be taken so that its viscosity is equivalent to the corresponding oil. Other forms of direct electronic sizing use a single charged probe which is connected to an electronic circuit which amplifies, classifies and counts the electronic pulses created by the droplets colliding with the probe.

A second rapid sizing technique utilises the scattering or attenuation of a light beam passed through the spray. The simplest application is to measure the light transmission through a known section of the spray. This technique readily gives a mean droplet size for a heterosized spray but gives no information on the droplet size distribution. An account of this technique has been given by Putnam and Thomas (1957).

2.2   MEASUREMENTS OF DROPLET VELOCITY AND DIRECTION

The most direct application of this technique is a variant of the direct photography technique but in which a double spark illumination technique is used. In this case the opening of the camera shutter is used to trigger one flash and then, after an appropriate interval (e.g. 10-100 μs) determined by an electronic time delay unit, the second flash unit discharges, after which the camera shutter closes. The resulting photograph now consists of a series of double images, each pair corresponding to each droplet. From the separation of each pair of images and from the knowledge of the time interval, the droplet velocity and direction may be deduced. In a more sophisticated application of this method, a high intensity stroboscopic light source may be synchronised with a cine camera to give additional information about droplet trajectories. Information on droplet velocities may also be deduced by laser-anemometer techniques in which a moving droplet influences the optical properties of two crossed (split) laser beams. The data obtained on the droplet velocity relate entirely to the small volume where the two split beams intersect.

2.3   DETERMINATION OF THE SPATIAL DISTRIBUTION OF DROPLETS PRODUCED BY AN ATOMISER

In addition to the knowledge of the drop size distribution and their velocities and direction in flight, an overall measurement of the spatial distribution of droplets is of practical importance. The spray pattern is a good indication of the precision of manufacture of a particular atomiser and for design purposes is of significance in assessing its behaviour in relation to combustion efficiency and performance.

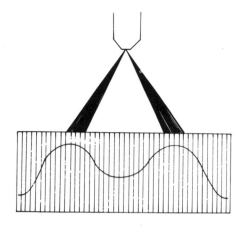

Figure 2.2. Spray distribution sampler (patternator).

Experimentally the spray distribution is most readily obtained by a 'patternator' of the type illustrated in Figure 2.2.   This device is illustrated in a form which gives the diametric variation of droplet concentration in a spray, the particular example relating to a hollow cone spray.   Variants of this are capable of giving the spatial variation across the whole of the spray.   It should be noted that visual or direct photographic techniques are capable of giving information about the periphery of the spray but are not capable of providing information as to the behaviour in the central regions of the spray.

2.4   MATHEMATICAL REPRESENTATION OF DROPLET SIZES

A number of mathematical expressions have been developed to express the mean droplet sizes and droplet size distributions of sprays.   It is not possible to relate the size of droplets produced by a particular atomiser with any theoretical analysis of droplet formation; consequently, droplet size distributions are represented by purely empirical expressions.

The size distributions of a spray may be represented in a number of ways, thus:

(a) the incremental number, ΔN, of droplets within the size range
$(d - \Delta d/2) < d < (d + \Delta d/2)$,

(b) the incremental volume (or mass), ΔV, of droplets in this size range,

(c) the cumulative number of droplets, N, less than a given size, and

(d) the cumulative volume (or mass) of droplets, V, less than a given
size d.

Typical forms of these quantities are plotted in Figure 2.3.

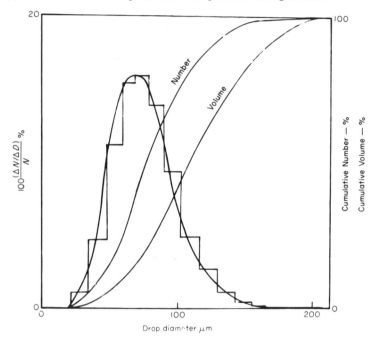

Figure 2.3.   Spatial drop size distribution and
cumulative number and volume.

The droplet size distribution of a spray may be represented by a
relatively simple mathematical function, namely

$$\frac{dN}{dd} = a d^{\alpha} \exp(-b d^{\beta}) \qquad (2.1)$$

where dN is the number of droplets which lie between the diameters d and
$d + dd$ and a, b, α and β are constants.   It has been shown that in
general α and β have the following values: $\alpha = -0.5$, $\beta = 1$ for a swirl
atomiser; $\alpha = 1$ and $\beta = 1$ for an impinging jet atomiser; $\alpha = 2$, $\beta = 1$
for a pressure jet atomiser, and $\alpha = 1$ and $\beta = 1$ for a twin-fluid

atomiser.   The parameters a and b are readily obtained by a graphical
method from experimental data for N and d by using Equation 2.1 in
the form

$$\log \frac{1}{d^\alpha} \frac{dN}{dd} = \log a - bd^\beta \log 1 \qquad (2.2)$$

and by plotting $(\frac{1}{d^\alpha} \frac{dN}{dd})$ against d^β.   A typical plot is shown in

Figure 2.4.

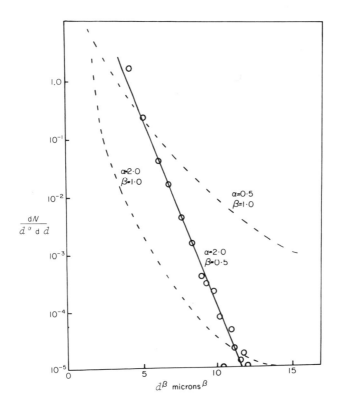

*Figure 2.4.   Plot of log (dN/ddd$^\alpha$) against d^β for droplets
produced by an oxy-oil atomiser where the
relationship is linear for α = 2, β = ½,
other assumed values of α and β produce curves
as shown.*

Now since $\quad V = \dfrac{N\pi d^3}{6}$

then $\qquad \dfrac{dN}{dd} = \dfrac{6}{\pi d^3}\ \dfrac{dV}{dd}$ $\hspace{4cm}$ (2.3)

Equation 2.1 can then be written in the form involving droplet volumes thus:

$$\frac{dV}{dd} = \frac{\pi d^3}{6}\ a\ d^\alpha\ \exp(-bd^\beta).$$ $\hspace{3cm}$ (2.4)

Equations 2.1 and 2.5 can also be integrated to give

$$N = \frac{a}{\beta b^{\alpha/\beta}}\left[\frac{\Gamma b\ d^\beta(\underline{\alpha+1})}{\beta} - \frac{\Gamma b d_o{}^\beta(\underline{\alpha+1})}{\beta} \right]$$ $\hspace{1.5cm}$ (2.5)

$$V = \frac{\pi a}{6\beta b^{\alpha+1}}\left[\frac{\Gamma b d^\beta(\underline{\alpha+1})}{\beta} - \frac{\Gamma b d_o{}^\beta(\underline{\alpha+1})}{\beta} \right]$$ $\hspace{1.5cm}$ (2.5^1)

where $\Gamma(\alpha)$ is the incomplete gamma function as defined by

$$\Gamma\alpha(n) = \int_0^\alpha \exp^{-x}\ x^{n-1} dx$$

Tables of gamma functions available and a short table of interest to spray systems is given in Appendix 3.

The Nukiyama-Tanasawa and Rosin-Rammler equations are special cases of the general equations. The Nukiyama-Tanasawa equation is similar to Equation 2.1 in that $\alpha = 2$ and the exponent β varies little from unity. Thus

$$\frac{dN}{dd} = ax^2\ \exp(-bd^\beta)$$ $\hspace{3cm}$ (2.6)

which by means of Equation 2.4 may be written in terms of volumes

$$\frac{dV}{dd} = \frac{\pi a}{6}\ d^5\ \exp(-bd^\beta)$$ $\hspace{3cm}$ (2.7)

The Rosin-Rammler equation was originally developed to express the size distribution of pulverised fuels and is usually expressed in the form

$$V = 1 - \exp(-bd^n) = 1 - \frac{R}{100}$$ $\hspace{3cm}$ (2.8)

where R is the percentage residue left on the sieves if a pulverised fuel is sized. Likewise, particles from a frozen wax sizing of liquid droplets would present experimental data of the same form.

Differentiation of Equation 2.8 gives

$$\frac{dN}{dd} = \frac{6bn}{\pi} d^{(n-4)} \exp(-bd^n) \qquad (2.9)$$

which is identifiable with Equation 2.1.

If the logarithm of Equation 2.8 is taken twice, the following equation linear in n and $\log b$ is obtained, i.e.

$$\log \log \left(\frac{100}{R}\right) = \log b + n \log d + \log \log e \qquad (2.10)$$

Thus a plot of $\log \log \frac{100}{R}$ against $\log d$ yields the value of n from its slope. This type of graph is called the Bennett diagram.

Another approach to the problem of size distribution has resulted from the application of a statistical analysis of the breakup of liquids and has led to the logarithmic-normal equation. This size distribution is based on an expression of the following type:

$$\frac{dN}{dd} = \frac{6}{\pi \sqrt{2\pi} \sigma d^4} \exp \left[- \frac{1}{2\sigma^2} \ln \frac{d}{M} \right]^2 \qquad (2.11)$$

Where M is the median for the logarithmic-normal distribution and σ is the standard deviation. Whilst this method is useful for a number of applications in spray characterisation it is difficult to apply to spray combustion systems and has not been widely used except in complex computer models.

Although a drop size distribution equation completely defining a spray is of most general application it is frequently convenient to express the drop size in the form of a mean diameter. Such a diameter can be calculated from size distribution equations or it can be obtained directly from the experimental data.

2.5 MATHEMATICAL REPRESENTATION OF MEAN DROPLET DIAMETERS

Useful average diameters are the mass median diameter (MMD) which is the diameter below or above which lies 50% of the mass of the droplets. The number median diameter (NMD) is the equivalent definition with reference to the number of drops. Both of these quantities may be obtained directly from the 50% point of the cumulative volume and number distribution curves respectively (Figure 2.3).

Mean diameters may be based on total number, surface or volume of drops in a spray. One of the most common mean diameters is the linear mean diameter, d_{10}, where

$$d_{10} = \frac{\int_0^\infty d \; dN}{\int_0^\infty dN} \qquad (2.12)$$

which strictly should be written thus

$$d_{10} = \frac{\int_{d_1}^{d_2} d \; dN}{\int_{d_1}^{d_2} dN} \qquad (2.13)$$

where d_1 and d_2 are the practical lower and upper limits. The concept of mean size has been generalised into one equation, namely

$$d_{qp} = \left[\frac{\int_0^\infty d^q \; dN}{\int_0^\infty d^p \; dN} \right]^{(q-p)^{-1}} \qquad (2.14)$$

where q and p take the values 1, 0 respectively for the linear mean diameter as in the example above, 2, 0 for the surface mean diameter, 3, 0 for the volume mean diameter, 3, 1 for the volume-diameter (mean evaporative diameter) and 3, 2 for the volume-surface or Sauter mean diameter. These quantities vary markedly. Thus, using Houghton's (1950) data for a particular set of droplets, it is possible to show (Dombrowski and Munday, 1970) that d_{10} = 5.5, d_{20} = 7.5, d_{31} = 13.5, d_{32} = 18, NMD = 4.2 and MMD = 24 µm.

A general expression for mean sizes may be derived from Equation 2.14 and from the generalised size distribution Equation 2.1; this is:

$$d \frac{q-p}{q-p} = \frac{\int_0^\infty d^q \left\{ ad^\alpha \; \exp(-bd^\beta) \right\} dd}{\int_0^\infty d^p \left\{ ad^\alpha \; \exp(-bd^\beta) \right\} dd} \qquad (2.15)$$

which may be solved using Equation 2.7. Thus, for example, we can obtain for the volume-surface mean diameter:

$$d_{32} = \int_o^\infty d^3 dN \bigg/ \int_o^\infty d^2 dN = b^{-1/\beta}\ \frac{\Gamma\{(\alpha + 4)/\beta\}}{\Gamma\{(\alpha + 3)/\beta\}} \quad (2.16)$$

whilst the quantities $\Gamma\{(\alpha + 4)/\beta\}$ etc. may be determined by means of Appendix 3.

2.6 DROPLET BALLISTICS

An understanding of the movement of droplets from an atomiser or in a combustion chamber is very important in any detailed design considerations.   This is particularly significant in relation to flame shape and as to whether the spray impinges on the combustion chamber wall.

The equation of motion of a single spherical particle, such as a droplet, discharged into a still gaseous medium with a velocity, V, is given by

$$m\ \frac{dV}{dt} + R = 0 \quad\quad\quad (2.17)$$

where m is the mass of the droplet and R is the resistance of the gaseous medium.   Gravitational forces may usually be neglected.

For a spherical droplet of constant mass, that is not undergoing evaporation, this may be rewritten thus

$$\rho_L\ \frac{\pi d^3}{6}\ \frac{dV}{dt} = C_D\ \frac{\pi d^2}{8}\ \rho_g\ (V - U)^2 \quad\quad (2.18)$$

or

$$\frac{dV}{dt} = \frac{3C_D \rho_g\ (V - U)^2}{4\ \rho_L\ d} \quad\quad\quad (2.19)$$

where ρ_L and ρ_g are the densities of the liquid droplet and gas respectively, C_D is the drag coefficient and V and U are the velocities of the droplet and the air respectively.

For the case of a spherical non-evaporating droplet at low Reynolds number, Stokes's law applies and the drag coefficient is given by

$$C_D = 24\ /\mathrm{Re} \quad\quad\quad (2.20)$$

Since the Reynolds number, Re, is given by

$$\mathrm{Re} = \rho_g\ Vd\ /\ \eta_g \quad\quad\quad (2.21)$$

where η_g is the viscosity of the gas.   Thus Equation 2.19 may be rewritten assuming laminar flow and a stagnant atmosphere thus:

$$\frac{dV}{dt} = \frac{-18\eta_g V}{d^2 \rho_L} \tag{2.22}$$

whilst under turbulent conditions, where $C_D = 0.44$, then

$$\frac{dV}{dt} = \frac{-0.33\rho_g V^2}{d\rho_L} \tag{2.23}$$

The velocity at time t, V_t, and the penetration distance, s, given by $ds/dt = V$, may be simply obtained by integration. Thus for laminar flow into a stagnant atmosphere

$$V_t = V_o \exp\left(\frac{18\eta_g t}{d^2 \rho_L}\right) \tag{2.24}$$

and

$$s = \frac{d^2 \rho_L}{18\eta_g} V_o \left(1 - \exp\frac{18\eta_g t}{d^2 \rho_L}\right) \tag{2.25}$$

where V_o refers to the initial (injection) velocity of the droplet. For turbulent flow the following pair of relationships is obtained:

$$V_t = V_o \left\{1 + (0.33\rho_g V_o t/d\rho_L)\right\}^{-1} \tag{2.26}$$

and

$$s = \frac{d\rho_L}{0.33\rho_g} \ln\left(1 + 0.33\rho_g V_o t/d\rho_L\right) \tag{2.27}$$

The drag coefficient for burning droplets varies in a fairly complex way with Re and this is discussed further in Section 4.6.

3 The Atomisation of Fuel Oils

In spray combustion it is necessary to atomise and distribute the liquid fuel in a controlled manner within the combustion chamber. The performance of the combustion unit is critically dependent upon the drop size produced by the atomiser and the manner in which the combustion air mixes with the droplets. In this chapter the process of atomisation and the means by which it is brought about will be considered. A detailed discussion of the mixing processes and the way the combustion process interacts with the atomisation process are discussed later in Chapter 6.

3.1 THE PROCESS OF ATOMISATION

Atomisers produce a fine spray of liquid by increasing the surface area of the liquid until it becomes unstable and breaks down, as illustrated in Plate 1.1. The exact mechanism is dependent upon the particular form of atomiser being considered and the nature of the liquid being atomised, but the basic mechanism involves the formation of unstable columns of liquid which break down into rows ot droplets. The process follows the mechanism suggested by Lord Rayleigh in which a column of liquid is unstable if its length is greater than its circumference. The column will then break down into a series of mono-sized droplets separated by one or more smaller satellite drops. However, because of the irregular character of the process, non-uniform threads are produced which results in a wide range of drop sizes.

Dombrowski and Munday (1968) have discussed in detail the mechanism of disintegration and have identified three modes of disintegration, namely, wave, rim and perforated sheet. The mechanism of sheet break-up, which is the case illustrated in Plate 1.1, may be idealised as in Figure 3.1. Here the waves on the sheet continue to grow until the crests are blown out and the sheet is broken up into half wavelengths which first form ligaments and then droplets.

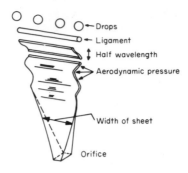

Figure 3.1.   *Idealised model of drop formation from a liquid sheet (after Dombrowski and Munday, 1968).*

The process of atomisation can be accomplished in a number of ways in practice.   The methods are usually grouped according to the source of energy used: (a) by forcing the liquid at high pressure through an orifice as in pressure jet atomisers, (b) by passing a stream of gas at high velocity over the liquid surface so that waves are generated which become extended into thin films as in the twin-fluid atomiser, (c) by impinging one liquid jet upon another liquid jet, (d) by the use of centrifugal forces as in the rotating cup atomiser, and (e) by the use of ultrasonic waves, which are propagated through the liquid. Some devices may use a combination of the techniques, as in hybrid atomisers.

3.2   PRESSURE JET ATOMISERS

This technique, which was developed in 1902, depends upon forcing the liquid through an orifice under pressure to form an unstable jet of high velocity which disintegrates after leaving the orifice.   An important feature of pressure atomisers is that the flow rate is proportional to the square root of the pressure, the square root law, so that for a given nozzle the following relationship holds:

$$\frac{\text{Flow rate}}{(\text{Pressure differential})^{\frac{1}{2}}} = \text{const (flow number)} \qquad (3.1)$$

The flow number (F.N.) is a characteristic parameter of any pressure atomiser.   Since a certain critical pressure must be attained before the atomisation process occurs this limits the lower flow rate that a particular atomiser can successfully handle.   This restricts the turn down ratio of the atomiser (the ratio of maximum to minimum oil flows)

which can be as low as 1.3:1 in a simple atomiser.   The upper limit is
normally limited by the fuel supply system.

Simple burners such as these are used as on-off burners in furnaces,
etc. in which the flow rate is not controlled over any significant flow
range.   They are also widely used in diesel engines.   A variety of
designs is available to increase the operating range up to approximately
10:1 and such 'wide range' burners may be subdivided into the classes of
spill atomisers, controlled oil recirculation and variable nozzles, the
latter incorporating a moving piston.

Generally oil is injected with a swirling motion (Swirl or Simplex
atomisers) and this is achieved by injecting the oil through tangential
slots into a swirl chamber where it rotates at high speed and escapes as
a spray.   A typical pressure jet nozzle incorporating swirl is shown in
Figure 3.2(a).   In this case the hollow conical sheet that is produced
emerges from the orifice with a tangential velocity which is
sufficiently high to cause an air core throughout the nozzle, producing
a hollow cone spray.   In full cone atomisers a further jet of liquid
enters the swirl chamber via an orifice so that the orifice is
continuously full.   Most atomisers used in practical applications are
of the conical type although flat sprays, fan sprays, of the type
illustrated in Plate 1.1, find applications in certain specialist
situations, such as aircraft afterburners.

In the spill or return flow atomisers, the fall-off in atomisation
performance of the pressure jet atomiser at low combustion rates is
prevented somewhat by maintaining a rapid flow of oil in the swirl
chamber by recirculating the oil back to the pump.   Two common
arrangements by which this is achieved are shown in Figures 3.2(b) and
3.2(c).   Generally fairly sophisticated fuel handling systems are
required to achieve maximum performance.   In gas turbine practice a wide
turndown range may be attained from atomisers employing a main and
subsidiary fuel system, each feeding two independent orifices, one much
smaller than the other.   The smaller orifice handles the lower flow
range and the larger orifice deals with the higher flows at higher
pressures.   These are the Duple and Duplex burners and a typical system
is illustrated in Figure 3.2(d).

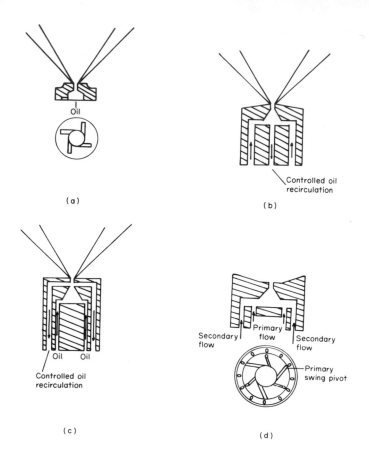

Figure 3.2. *Pressure jet atomiser types. (a) A simple swirl pressure jet atomiser, (b) a back spill atomiser, (c) a front spill atomiser, and (d) a duplex atomiser.*

In general pressure jet atomisers are available in a very wide range of capacities. When used with heavy fuel oils, the oil is preheated electrically or by steam tracing to about 15 to 24 cSt (70-100 s Redwood I) with oil pressures varying from 14 to 70 bar (200-1000 psig). The droplet sizes produced by the atomisers have been discussed by Dombrowski and Munday (1968) and a number of empirical equations are available depending upon the atomiser type and operating conditions. In the case of hollow core swirl atomisers it has been shown (Weinberg (1953)) that:

$$d_{32} \propto \left(\frac{FN^3}{\Delta P} \right)^{1/7} \tag{3.2}$$

In practice, Sauter mean droplet diameters are in the range 90 to 200 μm. Droplets produced by full core atomisers tend to be smaller than those in hollow core nozzles.

3.3 ASSISTED PRESSURE JET ATOMISER

This hybrid burner combines some features of both the pressure jet atomiser and the twin fluid atomiser. A typical atomiser is shown in Figure 3.3(a).

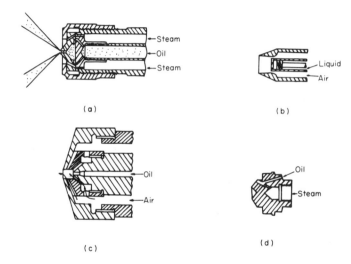

Figure 3.3. *Twin fluid atomiser types. (a) Assisted pressure atomiser, (b) simple low pressure air blast atomiser, (c) medium pressure air (MPA) atomiser, and (d) 'Y' type atomiser.*

It consists basically of a pressure jet tip with a series of atomising parts arranged to inject air or steam into the early stages of the fuel spray cone. These atomisers are of particular value to applications where large fuel throughputs per atomiser are required as in power station practice (in that case a single atomiser may burn 5 tonnes of oil per hour). The turndown ratio is typically 5:1. Fuel oil viscosities are similar to those for pressure jet atomisers, although the oil pressure is somewhat lower. The performance is essentially that of a pressure jet atomiser but the injected air or steam provides supplementary atomisation so that the maximum droplet size is reduced. This is of particular advantage in the large throughput burners because of the consequent reduction in smoke and grit emissions as further discussed in Chapter 7.

3.4   TWIN-FLUID (BLAST) ATOMISERS

In twin-fluid atomisers, a high velocity gas stream impinges on a relatively slow-moving liquid stream as in a scent spray. There are three main types:

    (a) low pressure, utilising air at less than 20 kN m^{-2} (3 psig) as the atomising medium,

    (b) medium pressure atomisers utilising air (MPA) at pressures between 20 and 100 kN m^{-2} (3-15 psig),

    (c) high pressure atomisers (HPA) normally using steam, at pressures exceeding 100 kN m^{-2}.

Atomisation may take place inside the burner or outside. In the former case, the steam and oil impinge within the burner and issue as a spray. In outside mix burners the oil is released into the gas stream at the outlet from the burner. Only about 3 to 10% of the air necessary for combustion is required for atomisation at medium or at high pressures, whilst 25% of the total air is required for atomisation at low pressures. Typical twin-fluid atomisers are illustrated in Figure 3.3(b) to (d).

The main advantage of twin-fluid atomisers is that they produce smaller droplets than can be achieved by means of a pressure jet atomiser. Furthermore, they are more intimately mixed with the combustion air than are pressure jet atomisers. However, the air requirements are more severe.

In the case of low pressure atomisers, in which a considerable part of the combustion air passes through the atomiser itself, the turndown ratio is restricted to the range of 2:1 to 5:1. Generally the fuel oil used (e.g. gas oil) has a viscosity of not greater than 15.5 cSt (70 s Redwood No.1) and is supplied by suction or under a slight gravity head.

In the case of a medium pressure air atomiser (MPA atomiser), a greater turndown ratio is possible, since less combustion air is used for atomisation. Ratios up to 10:1 can be obtained without impairing the efficiency of atomisation although a 4:1 turndown is more usual.

Fuel viscosities and feed pressures are similar to the LPA atomiser.

In the case of high pressure atomisers, steam is frequently used as the atomising fluid under pressures ranging from 2 to 7 bar (30-100 psig) since their applications are in situations such as steam raising where high pressure steam is available. Normally the turndown ratios may range from 5:1 to 15:1. The fuel oil is usually maintained at a viscosity of 30-37 cSt (120-150 s Redwood No.1). Oil pressures may vary from 6 to 20 bar, the latter being for a 15 or 20:1 turndown as is the case in marine applications.

The droplet sizes produced by these atomisers depend upon the atomiser design, the flow rates of fuel and air and the properties of the fuel. A number of correlations have been proposed and these have been discussed by Dombrowski and Munday (1968). Generally, atomisation in such a system is fine, with Sauter mean diameters of 40 - 80 μm, becoming finer with increasing gas velocity. In particular, the use of an internal mixing arrangement so that steam and oil mix to form an emulsion which expands as it issues from the final orifice results in a finely atomised spray of oil.

3.5   ROTARY ATOMISERS

In a rotary atomiser, oil is fed on to a rotating surface and atomisation of the oil is achieved by centrifugal force. The oil may be fed by means of a pipe on to a spinning disc or to the inner surface of a rotating hollow cup; both of these types of atomiser are illustrated in Figure 3.4(a) and (b).

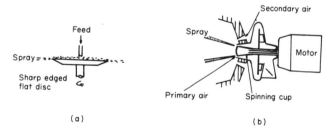

(a)   (b)

Figure 3.4.   *Rotary atomiser types.   (a) Sharp-edged disc of the type used for research applications, (b) air-blast spinning cup atomiser of the type used in industrial burners.*

The spinning disc atomiser has largely been used for research purposes or special applications and most studies relating to the mechanism of droplet formation have been undertaken with it. The

37

mechanism of droplet formation involves the spreading of the liquid
sheet on the disc and the formation of threads of liquid at the outer
edge of the disc. These threads then break into droplets. The length
and nature of the threads produced depend upon the liquid flow rate and
speed of rotation (Dombrowski and Munday, 1968). Because of the nature
of the breakup mechanism the droplets are mainly mono-sized and are
formed together with a small number of small diameter (satellite)
droplets.

If atomisation occurs when the accumulated mass of the droplet size
becomes greater than the retaining surface tension force then, as shown
by Walton and Prewett (1949),

$$\rho_L d^3 \times \omega^3 D \propto T d \qquad (3.3)$$

$$\text{or} \quad \omega d \left(\frac{D \rho_L}{T} \right)^{\frac{1}{2}} = \text{const} \qquad (3.4)$$

where ω is the disc angular velocity, D the disc diameter and T the
surface tension of the liquid.

Rotary atomisers of the cup type (Figure 3.4(b)) have a widespread
commercial application in residual fuel oil combustion. The cup is
rotated at high speed by means of an electric motor or by an air
turbine driven by a part of the atomising air. Typically, the cup
would rotate at 4000-6000 rpm to atomise an oil with a viscosity of
about cSt (400 s Redwood No.1). The air flow is directed so that the
droplets are projected forward rather than outwards, the atomiser in
practice behaving rather like a twin-fluid atomiser in which the
rotating cup prefilms the liquid to a sheet and the high velocity air
blast close to the cup-edge completes atomisation. The turndown ratio
for such atomisers is about 2:1 and, typically, the atomiser requires
that some 15% of the air required for combustion is supplied as primary
air around the cup. Dombrowski and Munday (1968) have listed the
equations correlating droplet size and operating parameters and shown
that the mean droplet diameter, d_m, is given by:

$$d_m = \frac{\text{const } \gamma^{0.33} W_S^{0.33}}{\rho_g^{0.17} \rho_L^{0.5} n D_L^{0.67}} \qquad (3.5)$$

Here γ is the surface tension, W_S is the liquid feed rate, ρ_g and ρ_L
the gas and liquid densities respectively, n the speed of rotation and
D_L the cup diameter. Typically, the diameters of droplets produced
would be about 200 μm at 4000 rpm for a fuel oil , decreasing to about

40 µm at 50 000 rpm.   Rotary atomisers are thus not very sensitive to
the viscosity of the fuel.

3.6   IMPINGING JET ATOMISERS

This type of atomiser, which is widely used in liquid-fuelled
rocket engines, is a special jet or pressure jet atomiser in which two
jets of liquid impinge resulting in atomisation.   Two types are
illustrated in Figure 3.5(a) and (b) representing liquid fuel-liquid
fuel impingement and liquid fuel-liquid oxidant impingement.   Typical
fuel-oxidant combinations are kerosine or liquid hydrogen with liquid
oxygen and hydrazine (or a derivative) with HNO_3 or N_2O_4.   In the case
of dissimilar liquids, mixing of the two liquids occurs prior to
atomisation.   Generally this type of atomiser produces smaller droplets
than the pressure jet atomiser and also produces less of the larger
droplets.

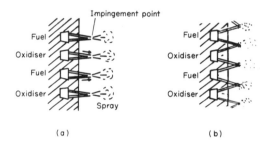

Figure 3.5.   *Impinging jet atomisers.*
(a) Fuel-fuel type, and
(b) fuel-oxidant type.

3.7   ULTRASONIC ATOMISERS

Two basic types of atomisers have been developed.   In the first,
atomisation is accomplished by actuating the atomiser surface by means
of a piezo-electric or magneto-strictive device.   Generally, the device
consists of a conical or flat vibrating surface on to which the fuel is
passed by means of a pipe.   Whilst such devices have received
considerable attention with respect to applications to small central
heating units, they are not used widely in practice.

The second method of ultrasonic atomisation involves the sonic
pressure wave generator nozzle.   A typical arrangement is shown in
Figure 3.6.   In this method a relatively low pressure gas, usually
compressed air or steam, passes at supersonic velocities through a

convergent-divergent nozzle into a resonator chamber and the resultant
high frequency pressure wave is focused into an open cavity.   The energy
in the acoustic wave atomises the fuel oil which is pumped via a pipe to
the cavity.   Generally, high frequencies are employed, varying between 6
to 100 kHz, and a relatively high degree of atomisation is achieved with
droplet sizes varying between 1 and 25 μm.   Typically, air pressures of
between 20 to 400 kN m^{-2} (3 to 60 psig) are required and typical air flows
are 7 dm^3 min^{-1} air per kg fuel (1 scfm per gallon).

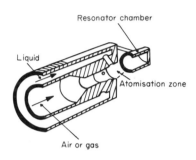

Figure 3.6.   *A resonator type sonic atomiser.*

3.8   OXY-OIL ATOMISERS

If the fuel oil is burned with oxygen or oxygen-enriched air,
special techniques have to be employed to prevent flashback of the flame
and explosion of the oxygen-oil mixture.   This can be achieved by the
use of two basic techniques which are illustrated in Figure 3.7(a) and
(b).

In the Toroidal burner, developed by Shell and shown in Figure
3.7(b), the oil is injected into a high velocity oxygen stream which
atomises the oil whilst preventing flashback.   The flame is stabilised
in a toroidal vortex and the shape of the flame can be controlled by
means of an outer water-cooled jacket.

The second technique, shown in Figure 3.7(b), utilises post-burner
mixing of the oxygen and the oil, and it is essentially a variant of the
impinging jet type of atomiser.   Here atomisation and mixing takes
place outside the burner, thus minimising explosion risks.

Because of the very high temperatures reached in oxy-oil combustion, both these types of atomiser incorporate water cooling to prevent the atomiser melting.

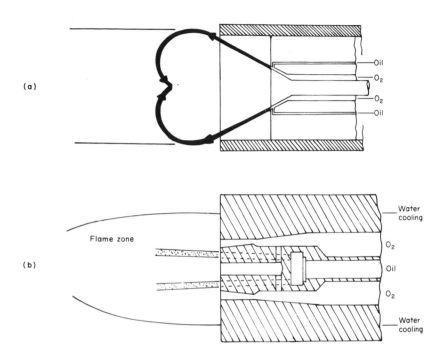

Figure 3.7.   Typical oxy-fuel burner types.   (a) Shell toroidal, and (b) post-face mixing (BOC Ltd).

41

4   The Combustion of Droplets of Liquid Fuels

4.1   THE NATURE OF SINGLE DROPLET COMBUSTION

The foundations of the present understanding of droplet combustion are based on work on the low temperature evaporation of droplets (mainly of water), but it was not until extensive work in the 1940s, prompted by the development of gas turbines and liquid-fuelled rockets, that the present day concept of droplet burning was developed.   Early work by Godsave and other workers (Godsave, 1953; Spalding, 1953) and which has been reviewed by Williams (1973) led to the classical spherico-symmetric diffusion-controlled model of combustion of a droplet in an oxidising atmosphere of the type which is illustrated in Figure 4.1.

Here, in the combustion of a liquid fuel in an oxidising atmosphere such as air, the liquid droplet evaporates and acts as a source of vapour.   Since the fuel vapour and air are initially separated, they burn in the form of a diffusion flame as illustrated in Plate 4.1. This particular photograph relates to a simulated droplet in which a porous sphere is used.   This technique, which is discussed later in the next section enables a detailed examination of the flame zone to be made.

Given such an understanding of droplet burning, it is then possible to regard a burning spray, of the type illustrated in Figure 1.1, as an ensemble of individual burning particles.   This, of course, is an idealised situation and, in practice, droplet combustion is usually more complex than this for two reasons.   Firstly, in dense sprays in which the amount of entrained air is small, droplets in the central parts of the spray may evaporate rather than burn and the vapour produced may burn at the outer boundary of the spray together with the droplets present at this boundary.   Secondly, in the case of the combustion of heavy fuel oils, their combustion may be more akin to the combustion of a coal particle, that is, in the early stages of combustion, extensive

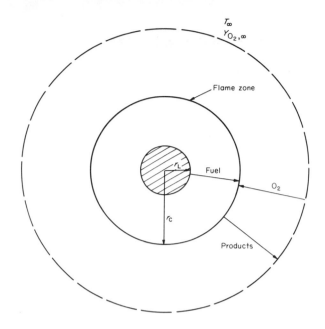

Figure 4.1.   Idealised model of droplet burning.

evaporation of the volatile components occurs.   These components burn
as a diffusion flame, but in the latter stages only a carbon-like residue
remains and this burns more slowly by a heterogeneous surface reaction in
the same way as a coke particle.

4.2   STUDIES OF THE RATE OF COMBUSTION OF VOLATILE DROPLETS

A considerable number of measurements have been made of the rate of
combustion of droplets of volatile liquids because the information
derived from such experiments has been of great importance to the
understanding of spray combustion in gas turbines and liquid fuelled
rockets.

The mass burning rate, $\dot{m}_F$, of a droplet is related to the rate of
decrease in droplet size by

$$\dot{m}_F = \frac{d}{dt}\left\{ \frac{4}{3}\pi r^3_L \rho_L \right\} \tag{4.1}$$

where m_F is the mass of the droplet, ρ_L is the density of the droplet at
the appropriate temperature and r_L is the droplet radius as defined in

43

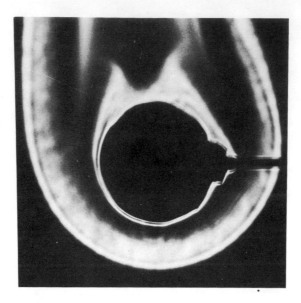

Plate 4.1. Typical photograph of single droplet combustion simulated by the porous sphere technique (see text).

Figure 4.1.   Equation 4.1 can be written in the form

$$\frac{d(d_L)^2}{dt} = \frac{2\dot{m}_F}{\pi \rho_L r_L} \tag{4.2}$$

where d_L is the diameter of the drop.

It has been shown many times experimentally, as discussed later, that under burning conditions the square of the droplet diameter is a function of the burning time, being related by a proportionality constant, K (or sometimes λ), which has been termed the burning-rate coefficient or evaporation constant.   It is a convenient parameter since it is readily obtained experimentally and is simply given by

$$-\frac{d(d_L)^2}{dt} = K. \tag{4.3}$$

This is frequently referred to as the 'd^2' law.   It is clear that by comparison of Equations 4.2 and 4.3

$$K \equiv -2\dot{m}_F / \pi \rho_L r_L.$$

4.2.1 Experimental Techniques

Three techniques have been used to investigate experimentally the rate of combustion of single droplets; these are: (i) the captive (or suspended) drop method, (ii) the supporting sphere technique and (iii) the free drop technique.

The captive drop technique may be used to obtain the rate of change of droplet diameter or size as a function of time. Typically, a single droplet (or array of droplets) is suspended from a silica fibre, ignited and the rate of combustion observed photographically as illustrated in Plate 4.2. From experiments of this type it is possible to obtain the burning-rate coefficient, K, from the 'd^2 law' using Equation 4.3 and typical experimental plots are shown in Figure 4.2.

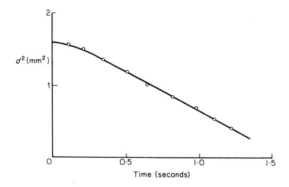

Figure 4.2. Typical plot of d^2 against time for combustion of a suspended droplet.

After an initial period in which the plots are non-linear (generally attributed to the establishment of steady-state conditions), the plots are linear, giving fairly accurate values for the burning-rate coefficient. In this manner, a considerable body of information has now become available on the burning-rate coefficients for a fairly wide range of fuels and various ambient conditions and these are discussed in the next section. A number of complications may arise during most studies. Firstly, the droplet is not spherical, due to the presence of the supporting fibre. Secondly, unless the experiments are conducted under zero-gravity conditions the flame shape is distorted and the mass burning-rate is enhanced by the influence of natural convection. Measurements of flame radius, r_c, or ratio of r_c to droplet radius, r_L,

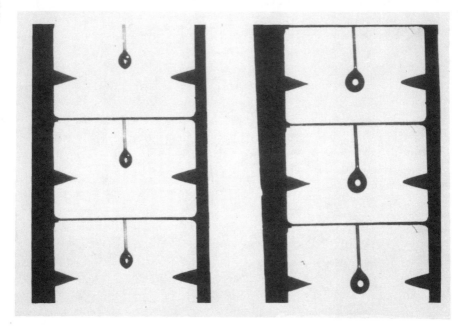

Plate 4.2.   Photographic records of the rate of droplet burning.   The photographs on the left hand side were taken some time after those on the right hand side.

may also be obtained by these techniques employing direct or schlieren photography.

Because of the simplicity of these measurements the technique has been applied to both bipropellants and monopropellants.   In the case of the former, the droplet is suspended in either a draught-free enclosure or in a furnace at the appropriate temperature and ignition is achieved by means of a spark, a pilot-light or self-ignition if a suitably high furnace temperature is used.   In the case of monopropellants the method of self-ignition is generally employed.

The supporting sphere technique provides a method of studying the steady-state combustion of simulated droplet burning. Here the diameter is kept constant during combustion by supplying fuel to the surface of a supporting inert sphere at a rate equal to the rate of its combustion. The technique involves the use of a ceramic porous sphere into which the fuel is fed by means of a small diameter stainless steel tube. The method is convenient in that different diameter support spheres may be used for a variety of experiments involving the steady-state combustion of droplets. In this way it has been used for measurements of flame shape, flame structure, various aerodynamic measurements and for measurements of mass-burning rates. It is the technique employed to obtain the photograph shown in Plate 4.1.

Ideally, studies of the behaviour of burning droplets actually in sprays would be invaluable but, with the exception of a few simple cases, the complexity of such systems prohibits the precise interpretation of data. Consequently, much data has been obtained by means of a variety of techniques using single droplets. In the free droplet technique, a single droplet or low density cloud of droplets is produced by a suitable generator, such as an ultrasonic atomiser, a vibrating steel tube, a spinning disc atomiser or a simple orifice such as a burette. The droplet of controlled size so produced is then allowed to fall under gravitational forces or is projected into a suitable hot environment such as a flame or furnace. In the latter case self-ignition occurs or is caused by a pilot flame. A typical arrangement is shown in Figure 4.3, in which a single droplet is dropped through a furnace. In some instances, a simple flame may be used as the source of the high temperature gases.

Generally, the extent of combustion is followed by direct photographic methods, although in the case of the combustion of heavy fuel oils the remaining carbon residues or cenospheres can be collected, weighed, and their structure examined by means of a microscope.

Investigations of the behaviour of droplets in high gas velocity gas streams may be undertaken by means of a shock tube. In this case the droplets are first dispersed inside the shock tube and combustion or droplet disintegration caused by the shock followed by means of an ultra high-speed photographic technique such as with an image camera.

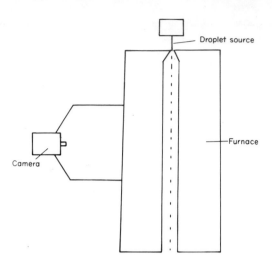

Figure 4.3.   *Diagrammatic representation of form of apparatus to
measure droplet burning rates of moving droplets.
This type of apparatus may also be used to measure
ignition delay times.*

4.2.2   <u>Experimental Data on Burning-Rate Coefficients</u>

A considerable amount of data has been obtained for single droplet
burning suspended on quartz fibres and this has been summarised by
Williams (1973).   Since experiments which are undertaken in the
laboratory are subjected to gravitational effects, in that the droplet
is not truly spherical, and that natural convective effects disturb the
flame, a number of measurements have been made under zero-convective
conditions.   In these experiments the quartz fibre with the droplet
suspended on it is contained in a box together with a cine camera which
is dropped from a sufficient height that essentially free-fall or zero
gravity conditions are realised.   In this way it has been found that
for n-heptane the burning-rate coefficient was 0.75 mm^{2}s^{-1} for a
suspended droplet and 0.79 mm^{2}s^{-1} for a free droplet, the small
difference being due to the conductive heat loss along the quartz
fibre.

Most experimental studies have been made of droplets burning under
normal gravitational forces, that is, influenced by natural convection,
and these measurements have been made for a number of fuels burning
under a variety of conditions.   For droplets of hydrocarbons burning

in air most fuels have burning-rate coefficients in the region of 0.8 to 1.2 mm^2s^{-1}. The values vary slightly with molecular weight. The variation of burning-rate coefficients of a series of n-alkanes are shown in Figure 4.4. Other useful values are as follows (units mm^2s^{-1}): cyclohexane 0.91; benzene 0.97; methyl alcohol 0.86; hydrazine 21.2: kerosene 0.96; diesel oil 0.79. A more comprehensive list has been given by Williams (1973).

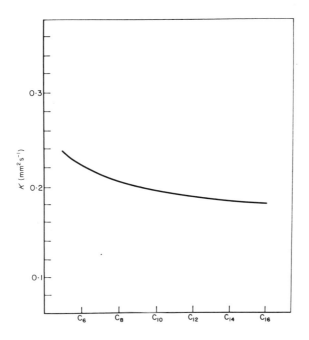

Figure 4.4.   Variation of burning-rate coefficients with chain length of droplets of n-alkanes burning in air.

A few measurements have been made with varying oxygen concentrations. Generally the burning-rate constants increase approximately linearly with oxygen content, the values for air being increased by a factor of 2 for 100% oxygen.

A considerable number of studies have been concerned with droplets burning in air at elevated temperatures, these experiments, of course, relating more closely to the circumstances found in furnace flames. The results obtained for a number of fuels are shown in Figure 4.5. It is clear that the higher molecular weight fuels have a higher temperature dependence, presumably due to enhanced radiant heat transfer to liquid phase cracking leading to carbon formation.   This is discussed further in Section 4.4.

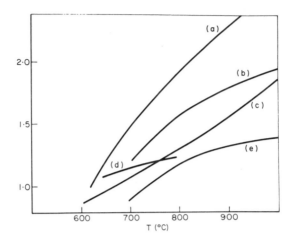

Figure 4.5. *Burning-rate coefficients of compounds burning in*
air at elevated temperatures.   (a) Cetane,
(b) commercial diesel fuel, (c) aviation kerosene,
(d)   n-heptane, and (e) benzene (after Williams, 1973).

A number of investigations of mass burning-rates have been made using the porous sphere technique.   These can be converted to equivalent burning-rate coefficients by means of Equation 4.3.   It should be noted, however, that values obtained in this way are not strictly equivalent to burning-rate constants obtained from suspended droplet techniques, because the heat conducted to the surface must both preheat the fuel to the surface temperature and also provide the heat of vaporisation.   This preheating requirement is not present during the steady-state burning of a suspended droplet once it has heated up.

By means of this technique, the variation of mass burning-rates has   been obtained under a variety of experimental conditions. Particularly the effects of natural convection and forced convection have been studied in this way.

In the case of spheres burning under conditions of natural convection it has been found that the combustion rate is a function of Grashof Number (Agoston *et al.* 1957).   A number of experimental investigations have been made of the influence of forced convection on

both burning and evaporating droplets and a number of empirical
equations have been postulated.   One of the most widely used is that by
Agoston *et al.* (1957) namely

$$\dot{m}_{F(forced)} = \dot{m}_o (1 + 0.24 Re^{\frac{1}{2}} Pr^{\frac{1}{3}}) \qquad (4.4)$$

where $\dot{m}_{F(forced)}$ and $\dot{m}_o$ are the mass burning-rates of the sphere under
forced convection and zero convection (zero gravity) conditions
respectively, and Re is the Reynolds Number.   The Prandtl Number, Pr,
was assumed to be unity.

4.3   THEORETICAL MODELS OF THE COMBUSTION OF VOLATILE DROPLETS BURNING
IN AIR

Theoretical models of droplet combustion are of great importance
since they give insight into the dependence of the rate of combustion on
the various chemical and physical processes that occur.   Furthermore,
they are of great value in the design or modelling of spray combustion
systems since it permits the direct *ab initio* calculation of the rate
of droplet burning.   Consequently, a number of theoretical
investigations have been undertaken over the past twenty-five years in
parallel with the experimental studies outlined in the previous sections.

The initial basic approach involved a spherico-symmetric model
based on a vaporising droplet in which the rate-controlling process is
molecular diffusion rather than chemical kinetic factors.   This model
has developed as a result of work by numerous investigators who have
proposed a number of approaches but which all tend to the same kind of
result (Williams, 1965).

Generally, it is assumed that the droplet is spherical and is a
pure liquid (few multicomponent analyses have been undertaken).   In
addition, the following assumptions are generally made in the basic
model:

(i)   The combustion system has spherical symmetry as shown in
Figure 4.1.   The spherical droplet of radius r_L is surrounded by
a concentric flame zone of radius r_c.   Concentric with the droplet
and outside the flame zone lies another outer boundary which is
taken to be at infinite distance from the droplet, and the
composition of which is that of the ambient atmosphere.

(ii)   The flame is assumed to be supported by the exothermic
reaction of fuel and oxygen in the flame zone, the oxidant diffusing
in from the outer boundary to the flame zone whilst the fuel vapour
diffuses from the droplet surface.   Heat is transferred from the

flame zone to the droplet to provide the latent heat of vaporisation of the liquid fuel; this not being necessary in the special case of combustion in critical conditions. As the critical temperature and pressure are approached the latent heat of vaporisation decreases to zero.

(iii) Thermal diffusion effects are neglected.

(iv) The effect of radiant heat transfer from the gas phase or adjacent particles is negligible; this has been shown to be a good assumption except in the case of heavy fuel oils.

The earlier analyses also made two other major assumptions, firstly, that combustion occurs under quasi-steady state conditions and, secondly, that the chemical reaction rate is infinitely fast. Analyses based on these assumptions have given the classical equations of droplet burning, but subsequently workers have questioned the validity of these assumptions.

4.3.1 Analyses Based on Quasi-Steady State Combustion and Infinite Kinetics

Here it was assumed that soon after ignition the droplet settled down to steady-state conditions, and indeed experimental plots of d^2 against t seemed to confirm this since they became linear after a short time interval. In addition the quasi-steady state theories also assume that the temperature in the interior of the liquid is invarient with time and close to the surface temperature. This surface temperature, T_L, is taken to be that of the boiling point of the liquid.

This assumption has been considered in detail and involves little error except in extreme cases.

To make any theoretical prediction of the burning-rate coefficient and temperature and composition field surrounding a burning droplet it is necessary to use the general continuity equations developed by Hirschfelder *et al.* (1954).

These equations take the general form:

Global mass conservation

$$\frac{\partial \rho}{\partial t} + \frac{1}{r^2} \frac{\partial r^2 \, v\rho}{\partial r} = 0 \qquad (4.5)$$

which reduced on integration, when $\partial\rho/\partial t = 0$, to

$$\dot{m}_F = 4\pi \, r^2 \, \rho v = \text{constant} \qquad (4.6)$$

where $\dot{m}_F$ is the mass flow of the fuel vapour leaving the surface (i.e. the mass burning rate), r is the radial co-ordinate, ρ the gas density and v the radial velocity.

Species continuity

$$\frac{\partial}{\partial t}(\rho Y_i) + \frac{1}{r^2}\frac{\partial r^2}{\partial r}\left\{\rho v Y_i - \rho D_i \frac{\partial Y_i}{\partial r}\right\} = q_i \qquad (4.7)$$

where Y_i is the weight fraction, D_i the diffusion coefficient and q_i the chemical rate of formation of species i. This may be rewritten as

$$\frac{\partial(\rho Y_i)}{\partial t} + \frac{\dot{m}_F}{4\pi r^2}\frac{\partial Y_i}{\partial r} - \rho D \frac{1}{r^2}\frac{\partial}{\partial r}\left\{r^2 \frac{\partial Y_i}{\partial r}\right\} = q_i \qquad (4.8)$$

Conservation of energy

For an adiabatic system this takes the form

$$\frac{\partial}{\partial t}\rho \Sigma_i (Y_i H_i) + \frac{1}{r^2}\frac{\partial r^2}{\partial r}\left\{\rho v \Sigma_i (Y_i H_i) - \lambda\frac{\partial T}{\partial r}\right\} = -\Sigma q_i H_i \quad (4.9)$$

where H_i is the enthalpy of species i, λ the thermal conductivity of the mixture and T the temperature. This may be rewritten so that the chemical rate of heat production is incorporated in the enthalpy terms and the radial velocity term replaced by $\dot{m}_F$ by means of Equation 4.6.

The equation of motion

$$\frac{\partial(\rho v)}{\partial t} + \frac{1}{r^2}\frac{\partial}{\partial r}(r^2 \rho v^2) = -\frac{\partial p}{\partial r} \qquad (4.10)$$

where p is the pressure. Normally it is assumed that combustion occurs under constant pressure.

In the quasi-steady state analysis the time dependent terms become zero and the boundary conditions are imposed on the basis of negligible liquid-phase reactions within the droplet and completion of chemical reaction at the outer boundary.
That is, at $r = r_L$, $T = T_L$ and $Y_F = Y_{F,L}$

and at $r = r_\infty$, $T = T_\infty$ and $Y_O = Y_{O,\infty}$

where Y_F and Y_O are the mass fractions of the fuel vapour and oxygen respectively.

The conservation equations are in an exact form but, because of the boundary conditions and the form of the expression for the chemical reaction rate, analytic solutions of the equations are not possible.

Consequently, the major assumption of an 'infinitely rapid chemical reaction rate' is made, that is, the chemical reaction rate is not controlling the rate of droplet disappearance in any way. Usually this assumption is coupled with the use of a simplified chemical reaction of the type

$$\text{Fuel} \quad + \quad \text{Oxygen} \quad \longrightarrow \quad \text{Products.}$$

In this model the flame zone, as a consequence of the infinite rate assumption, is of infinitesimal thickness and would thus be represented by a surface rather than an extended reaction zone, as indicated by Figure 4.6.

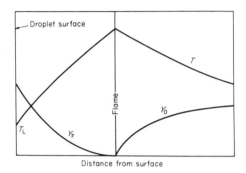

Figure 4.6. *Variation of temperature (T), fuel vapour (Y_F) and oxygen (Y_O) as a function of distance from the surface of a burning droplet. This representation corresponds to the 'thin flame' model.*

Generally, it is necessary to assume that the fuel and oxidant diffuse to the reaction zone in stoichiometric proportions and again, as a consequence of the infinite reaction rate assumption, their concentrations become zero at the reaction interface. When these approximations are made, an analytic solution may be obtained. The forms of the final equations depend upon the exact method of solution and, in particular, if fixed or variable transport properties are assumed.

A typical and widely used solution based on temperature-independent transport properties is that given by Wise and Agoston (1958):

$$\dot{m}_F \quad = \quad 4\pi \, r_L \quad \overline{\rho} \, \overline{D} \quad \ln \, (1 \quad + \quad B) \tag{4.11}$$

where $\overline{\rho}$ is the averaged gas density, $\overline{D}$ the averaged diffusivity and B is the transfer number given by

$$B \equiv 1/L \left\{ \overline{C}_p \ (T_\infty - T_L) + Q \ Y_{o,\infty/i} \right\} \quad (4.12)$$

which in the evaporative limit (i.e. evaporation in hot combustion products) becomes

$$B_{ev} \equiv \overline{C}_p \ (T_\infty - T_L)/L \quad (4.13)$$

where $\overline{C}_p$ is the average specific heat, T_∞ the ambient temperature, T_L the temperature at the liquid surface (usually taken as the boiling point of the liquid), Q the heat of combustion, L the latent heat of vaporisation/unit mass evaporating and i the stoichiometric mixture ration, (Y_o/Y_F) stoich.

The quantity $\overline{\rho} \ \overline{D}$ in Equation 4.11 is usually replaced, assuming a Lewis number (Le) of unity, by $\lambda/\overline{C}_p$. In which case, and using Equation 4.3, an expression is obtained for the burning-rate coefficient thus:

$$K = \frac{8 \ \overline{\lambda}}{\overline{C}_p \rho_L} \ \ell n \ (1 + B) \quad (4.14)$$

Similarly expressions have been derived for the flame temperature, the mass fraction of fuel vapour at the droplet surface $(Y_{F,L})$ and the ratio of flame radius to droplet radius (r_c/r_L). Thus Wise and Agoston obtained:

$$T_f - T_L = \frac{Q - L}{C_p} \left\{ \frac{1}{1 + i/Y_{o,\infty}} \right\} + \left\{ \frac{T \quad - T_L}{1 + Y_{o,\infty/i}} \right\} \quad (4.15)$$

$$Y_{F,L} = 1 - \left\{ (1 + Y_{o,\infty/i})/(1 + B) \right\} \quad (4.16)$$

$$\frac{r_c}{r_L} = \left\{ \ln \ (1 + B) \right\} / \left\{ \ln \ (1 + Y_{o,\infty/i}) \right\} \quad (4.17)$$

The simple spherico-symmetric flame model assumes that $\overline{\lambda}$ and $\overline{C}_p$ are constant and that the Lewis Number is unity. The results obtained thus markedly depend upon the choice of λ and C_p and since they are both functions of both temperature and composition almost any approach is certain to be unsatisfactory. Both λ and C_p are usually evaluated on the basis that the gas composition is air or nitrogen and their values calculated at the arithmetic mean of T_L and T_f, although, alternatively,

55

the logarithmic mean is often used.   More sophisticated approaches have
been adopted which have incorporated temperature dependent expressions
for λ and C_p.

4.3.2   Analyses based on Quasi-Steady State Combustion and Finite Kinetics

Several investigations have considered the solution of the
quasi-steady state equations without resort to the assumption of an
infinitely rapid reaction rate.   In general, analyses of this type have
yielded calculated burning-rate coefficients in reasonable agreement with
those obtained by the simple diffusion model and with the experimental
values, and thus have generally been taken as proof of the validity of
the diffusion model approach.   However, there are some considerable
limitations to the methods used since the rate of chemical reaction has
been assumed to be of the form

$$\text{Rate} = A \left[\text{Fuel}\right]\left[\text{Oxygen}\right] \exp (-E/RT)$$

where A and E are Arrhenius type reaction rate parameters.   It has been
shown that for typical hydrocarbon fuels burning under normal conditions
little error is introduced to the mass burning rate relationship by
assuming infinite chemical reaction rate except under extreme conditions,
i.e. extremes of pressure or droplet size.   Analyses based on finite
chemical reaction rates do however give more realistic indications of the
structure of the reaction zone.

4.3.3   Analyses Based on Non-Steady State (Transient) Combustion

Experimental work carried out by a number of researchers has
indicated that although the 'd^2 law' is fairly well obeyed throughout the
combustion of a droplet, the heat and mass transfer processes are in a
transient state during the major part of the droplet lifetime.   This is
manifest by small changes in r_c/r_L during the droplet burning and the
changes in the temperature within the droplet.   Theories of transient
combustion are complex and as yet in their infancy.   Any advances in its
understanding are unlikely to necessitate any revision to our
understanding of spray combustion as a whole but may yield valuable
information on the way droplets of heavy fuel oils are converted to
carbon particles.

4.4   THE COMBUSTION OF HEAVY FUEL OILS

Many commercial fuels consist of mixtures of hydrocarbons which have
widely differing boiling points.   Their behaviour when burning may thus

be markedly different from single component volatile liquid droplets in a number of respects. Firstly, selective distillation of the components may occur and if the lighter fractions are very volatile then these may flash evaporate forming bubbles within the droplet which disintegrate the droplet. This is termed 'disruptive' evaporation. The second characteristic is that the multicomponent combustion of mixtures containing high boiling point components may result in droplet temperatures that are so high that thermal decomposition occurs, resulting in the formation of coke-like residues termed cenospheres.

If a multicomponent mixture, such as a turbo-fuel, consists of reasonably similar components the 'd^2 law' is obeyed and the burning-rate constant for the mixture can be deduced from the burning-rate constants of the constituents by means of a simple additivity rule (Williams, 1973). If the components are moderately dissimilar, as in a gas oil, the 'd^2 law' is still reasonably obeyed but the rate of combustion is enhanced at high temperatures as shown in Figure 4.5.

Heavier fuel oils burn in a much more complex manner. Medium fuel oils may obey the 'd^2 law' during the early stages of combustion but such plots show an abrupt change when the cenosphere is formed and this burns at a much slower rate, about 1/10th of the rate of the droplet. In the case of heavy fuel oils, frothing of the droplet may occur in the fairly early stages of combustion due to disruptive boiling; this is followed by cracking of the residual material, resulting in the formation of cenospheres. Some typical cenospheres are shown in Plate 4.3 and the open structure should be noted.

Little is known about the mechanism of cenosphere formation under such circumstances except that the boiling points of the constituents and their chemical nature are both important (Williams, 1973).

4.5 COMBUSTION OF DROPLETS OF MONOPROPELLANTS

Monopropellants, such as hydrazine (N_2H_4) and ethyl nitrate ($C_2H_5NO_3$) which are used in rockets, can maintain a self-sustained flame without any additional oxidant. Consequently, monopropellant droplet combustion resembles a premixed spherical flame rather than diffusion-controlled droplet combustion, since the source of both fuel and oxidant is the droplet itself. Because of this, monopropellant combustion is much more complex than bipropellant combustion.

The burning-rate coefficient is not simply given by the 'd^2 law'

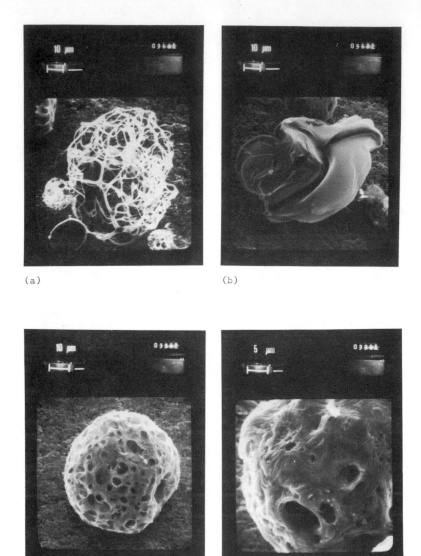

(a)

(b)

(c)

(d)

Plate 4.3. Residual fuel oil droplet residues.
(a) and (b) Characteristic skeletal and
membranous structures, (c) and (d) typical
porous coke spheres from an air heater hopper
in a power station (Street, 1974).

but follows a more general relationship thus

$$K = \frac{-d\,d^n}{dt} \tag{4.18}$$

where n varies from the square law dependence to unity depending upon
the type of monopropellant being burned and its droplet size.

Sometimes monopropellants are used as a fuel in a bipropellant
system, for example, hydrazine + dinitrogen tetroxide. In such systems
the main features of both types of system appear and the decomposition
region near the liquid surface would be surrounded by the oxidation
zone. The presence of both modes of combustion as illustrated in
Figure 4.7 is termed hybrid combustion.

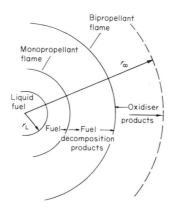

Figure 4.7. *Diagrammatic representation of model for the hybrid
combustion of a propellant (after Allison and Faeth, 1972).*

4.6 THE COMBUSTION OF MOVING DROPLETS

In order to be able to understand the characteristics of a burning
spray it is necessary to have information on the effects of the droplet
movement on the rate of combustion, and on the trajectory of the droplet.
The laws governing the mass burning rates of droplets have been outlined
previously (Equation 4.4) and this section is primarily concerned with
the drag coefficients of evaporating or burning droplets.

It is conventional to compare the drag coefficients of evaporating
or burning droplets with those of smooth inert spheres moving in
steady-state conditions, in which case the drag coefficient, C_D, is a
function of the approach Reynolds number as described previously in
Section 2.4. Burning and evaporating droplets differ from that ideal

model in that when the vapour effuses from the droplet surface the skin
drag decreases due to the thickened boundary layer.   In addition, there
may be changes in the nature of the flame, resulting from the relative
velocity of the droplet to the gas.   If the two velocities are similar
then the flame surrounding the burning droplet is near spherical, called
an envelope flame.   At higher velocities the drag forces cause the flame
to trail behind the droplet as a so called wake flame.   These two types
of flames are illustrated in Figure 4.8.   The type of flame obviously

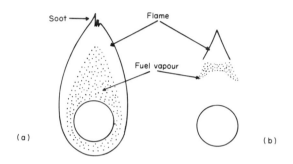

Figure 4.8.   *Flame types.   (a) Envelope flame, and*
(b) wake flame.

influences the drag coefficient as well as the change in velocity itself.
In view of the many complicating features the drag coefficients of
burning and evaporating droplets are not known with any precision.
However, they may be summarised with reasonable accuracy by the following
equations due to Dickerson and Schuman (1956).

$$C_D = 27 \ Re^{-0.84} \quad (0 < Re \leqslant 80)$$

$$C_D = 0.271 \ Re^{0.217} \quad (80 < Re \leqslant 10^4)$$

$$C_D = 2 \quad (10^4 < Re)$$

4.7   IGNITION OF DROPLETS

There are essentially two types of processes by which the ignition
of a droplet may be accomplished.   The first is thermal ignition, which
occurs when a cold droplet is exposed to a hot air or other oxidising
atmosphere, as in a diesel engine.   The second is the network mechanism,
which occurs when cold droplets are in contact with a burning spray and
ignition results from the combined influence of the transfer of heat and

of free radicals from the burning droplets to the approaching cold
droplets. This mechanism is important in stationary combustion
systems such as furnaces. In certain systems radiative heat transfer
to the droplet may play a role but generally it is unimportant.

4.7.1 Thermal or Spontaneous Ignition

This type of ignition proceeds via a series of steps which are
shown in Figure 4.9(a) and which are as follows:

(i) The surface temperature of the droplet increases primarily
   by conduction processes (but may be augmented by radiation)
   and the vapour pressure of the fuel increases.

(ii) The fuel vapour mixes with the oxygen in the air and the
   temperature of the fuel vapour-air mixture increases.

(iii) The composition of the mixture becomes within the
   inflammability limits for the particular fuel-air mixture
   and the temperature increases so that it exceeds the
   ignition temperature; at this point the rate of the
   hydrocarbon-oxygen chain reaction (Bradley, 1970) exceeds
   the critical limit and ignition occurs.

The time taken for ignition to occur is called the ignition delay,
the time relating to the occurrence of the physical processes,
vaporisation, mixing, etc. is termed the 'physical delay' and the time
relating to the chamical processes is termed the 'chemical delay'.
Obviously the processes overlap for part of the ignition period.

Ignition delay periods, t_i, may be represented by means of the
equation

$$t_i = A \exp \left(\frac{B}{T} \right)$$

where A and B are constants. B is often referred to as an overall
activation energy and has values between 6 and 80 kJ mol^{-1} depending
upon the nature of the hydrocarbon.

The magnitude of the ignition delay time, which is largely
determined by A, is very markedly dependent upon the chemical structure
of the hydrocarbon being ignited. This is largely due to the fact
that the onset of ignition is determined by the chemical reaction of
hydrocarbon vapour and oxygen, both in the gaseous phases. The ease
of oxidation of the hydrocarbon is important. Once ignition has
started the rate of combustion is largely independent of hydrocarbon
type since it is now controlled by a physical law, the 'd^2 law'. The
ignitability of hydrocarbon droplets by the thermal mechanism is most

(a) Spontaneous ignition

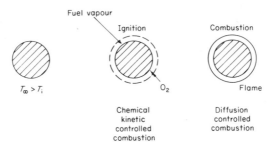

(b) Network mechanism of ignition

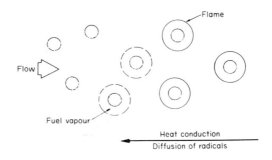

Figure 4.9. *Ignition of droplets.* *(a) Spontaneous mechanism, and (b) network mechanism.*

important in diesel engines and fuels with short ignition times, i.e. reactive hydrocarbons such as cetane (n-hexadecane, $C_{16}H_{34}$) have a high cetane number whilst fuels with a long ignition delay have a low cetane number. By definition, the standard reference fuels are cetane, with a cetane number of 100, and α-methyl naphthalene, with a cetane number of 0. The ignition delay periods may be significantly reduced by the addition of small quantities of 'promoters'. These may be alkyl peroxides, nitrites or, in practice, the more stable alkyl nitrates such as ethyl or *iso*-amyl nitrate. These compounds thermally decompose at relatively low temperatures thus facilitating the gas phase ignition step.

Ignition delay times for the ignition of single droplets may be
obtained by a variant of the suspended droplet experiment.   Here a
droplet of the appropriate liquid fuel is suspended on a silica fibre and
a furnace heated to some temperature above the ignition temperature is
rapidly positioned around it.   Usually the furnace is on a trolley and
when it is in the correct position it actuates a cine camera which
follows the course of ignition.   An alternative technique is to drop
single droplets into a furnace and measure the time interval by means of
a cine camera, such an apparatus is shown in Figure 4.3.

4.7.2   Network Ignition

This form of ignition, illustrated in Figure 4.9(b), is applicable
to spray situations.   As the cool droplets approach a flame front they
are heated by convection and radiation, and, furthermore, free radicals
also diffuse to them thus initiating the chemical reaction between the
hydrocarbon vapour and the oxygen.   Thus this form of ignition is not
very dependent upon the chemical composition of the fuel as is the case
in thermal ignition.   The application of network ignition to the
understanding of spray flame propagation rates is outlined in Chapter 5.

5 The Theoretical Modelling of Spray Combustion Systems

The theoretical modelling of the behaviour of a spray flame in a combustion chamber is of considerable importance in the design or improvement of combustion equipment, since the influence of many of the controlling parameters can be tested rapidly and more cheaply than testing prototype combustion chambers.

Mathematical analyses of spray combustion systems are frequently based on one-dimensional models of the type shown in Figure 1.1. Here the flame is considered to be essentially a flowing reaction system in which the time scale of the usual reaction rate expression is replaced by a distance scale.

As the unburnt spray approaches the flame front, it first passes through a region of preheating during which some vaporisation occurs. As the flame zone is reached the temperature rapidly rises and the drops burn. The flame zone can thus be considered to be a localised reaction zone sandwiched between, on one side, a cold mixture of fuel and oxidant, and on the other, the hot burnt gases. If the gas flow through the flame is one-dimensional along, for example an x-axis of co-ordinates, then the flame front is planar. The nature of the burnt products depends much upon the properties of the spray in its unburnt state. If the droplets are large then combustion may not be complete in the main reaction zone and unburnt fuel will penetrate well into the burnt gas region. If the droplets are small then a state of affairs exists that approximates very closely to the combustion of a premixed gaseous flame. Here the droplets are vaporised in the preheat zone and reaction after that is between the two reactants in their gaseous state. The other factors which determine the time to reach complete combustion, that is, the length of the combustion zone, are the volatility of the liquid fuel, the ratio of fuel to oxidant in the unburnt mixture and the uniformity of mixture distribution.

In most practical systems, such as a furnace or a rocket engine, the combustion process is much more complicated, due to turbulence or recirculation, and frequently cannot be represented by simplified models.

However, despite the complexity of this scheme, much valuable information has been obtained by the use of simplified analysis. In all these cases, a major problem that has been considered, and to a large extent solved, is concerned with the mechanism of combustion of the individual droplets that make up the spray.

To obtain a full understanding of the processes involved in spray combustion it was first necessary to have complete knowledge of

(a) the mechanism of combustion of the individual droplets that make up the spray,

(b) the description of the droplets that make up the spray with regard to size and spatial distribution, and

(c) any interaction between the individual droplets when they undergo combustion in the spray.

Of these, the mechanism of single droplet combustion is fairly well established as shown in Chapter 3. Furthermore, the statistical nature and properties of sprays are fairly well understood.

5.1   SPRAY COMBUSTION IN ONE-DIMENSIONAL SYSTEMS

The first theoretical model was put forward by Probert (1946) who calculated the influence of spray droplet size on the rate of combustion. However, Probert, and more recently Nuruzzaman *et al.* (1971), used analyses based on the Rosin-Rammler distribution law.

A more general procedure, applicable to a variety of droplet size distribution functions, has been developed by Tanasawa and Tesima (1958). Essentially the method requires knowledge of a droplet burning law, namely Equation 4.3 which may be integrated to give

$$d^2 = d_o^2 - Kt \qquad (5.1)$$

where d_o is the initial diameter of the droplet. This equation is particularly useful in relation to mono-sized sprays in that the diameter of droplets at time, t, is readily calculated from a knowledge of t, d_o and K. Furthermore, the time required for the droplet to be completely burned, the burn-out time, t_B, is obtained by setting $d = 0$ as in Equation 5.2,

$$t_B = d_o^2/K \qquad (5.2)$$

In heterosized sprays such calculations are not possible and a more complex analysis is necessary. The first requirement in such systems is knowledge of the droplet size distribution and here the law previously outlined is applicable, namely

$$\frac{dN}{dd} = ad^\alpha \exp{-bd^\beta} \qquad (2.1)$$

The volume-surface mean diameter is given by

$$d_{32} = b^{-1/\beta} \frac{\Gamma\{(\alpha + 4)/\beta\}}{\Gamma\{(\alpha + 3)/\beta\}} = \overline{d} \qquad (2.16)$$

and the total number of droplets by

$$n = \int_0^\infty dN = \frac{\alpha}{\beta b^{(\alpha+1)/\beta}} \Gamma \left(\frac{\alpha + 1}{\beta}\right)$$

so that

$$\frac{dN}{N} = \left[\frac{\Gamma\{(\alpha+4)/\beta\}}{\Gamma\{(\alpha+3)/\beta\}}\right]^{\alpha+3} \frac{\beta}{\Gamma\{(\alpha+1)/\beta\}} \left(\frac{d}{\overline{d}}\right)^\alpha \exp\left\{\left(\frac{\Gamma\{(\alpha+4)/\beta\}}{\Gamma\{(\alpha+3)/\beta\}}\right)^\beta \left(\frac{d}{\overline{d}}\right)^\beta\right\} \frac{dd}{\overline{d}} \quad (5.3)$$

the weight distribution, dw, is given by

$$dw = \rho_L \frac{\pi d^3}{6} dN = \rho_L \frac{\pi ad}{6}^{\alpha+3} \exp(-bx^\beta) dd \qquad (5.4)$$

and the total weight, w, of the droplets given by

$$w = \int_0^\infty \rho_L \frac{\pi d^3}{6} dN = \rho_L \frac{\pi}{6} \frac{\alpha}{\beta b^{(\alpha+4)/\beta}} \Gamma \frac{\alpha+4}{\beta} \qquad (5.5)$$

and thus

$$\frac{dw}{w} = \beta \frac{[\Gamma\{(\alpha+4)/\beta\}]^{\alpha+3}}{[\Gamma\{(\alpha+3)/\beta\}]^{\alpha+4}} \left(\frac{d}{\overline{d}}\right)^{\alpha+3} \exp\left[-\left\{\frac{\Gamma\{(\alpha+4)/\beta\}}{\Gamma\{(\alpha+3)/\beta\}}\right\}^\beta \left(\frac{d}{\overline{d}}\right)^\beta\right] \frac{d}{\overline{d}} \qquad (5.6)$$

Thus, the weight of the group of drops after combustion for time, t, and whose diameters lie between d and $d + dd$ becomes

$$dw_t = \frac{\pi \rho_L}{6} d^3 (dN)_t = \frac{\pi \rho_L}{6} d^3 (dN)_o \qquad (5.7)$$

By means of Equations 5.4 and 5.6 it can be shown that

$$\frac{(dw)_t}{d\xi} = \frac{\beta \left[\Gamma\{(\alpha+4)/\beta\}\right]^{\alpha+3}}{\left[\Gamma\{(\alpha+3)/\beta\}\right]^{\alpha+4}} \left(\xi_o^2 - \tau\right)^{3/2} \xi^\alpha \exp\left[-\left\{\frac{\Gamma\{(\alpha+4)/\beta\}}{\Gamma(\alpha+3)/\beta}\right\}^\beta \xi_o^\beta\right] \qquad (5.8)$$

where $\xi = d/\bar{d}$ and $\tau = Kt/\bar{d}^{-2}$

so that $\xi^2 = \xi_o^2 - \tau$

that is, a non-dimensional expression for the d^2 law.

Integration of Equation 5.8 gives the mass of droplets burned, w_b, thus

$$\frac{w_b}{w_o} = \frac{w_o - w_t}{w_o} = 1 - \int_{t_{\frac{1}{2}}}^{dmax/\bar{d}} (dw) \qquad (5.9)$$

These equations may then be readily applied to the combustion of droplets. Thus the change of weight distribution may be calculated by Equation 5.6 and this gives curves of the form shown in Figure 5.1. The amount of fuel consumed, in terms of mass, may be calculated by means of Equation 5.9 and typical curves obtained in this way are given in Figure 5.2.

5.2 COMBUSTION IN ROCKET COMBUSTION CHAMBERS

If combustion takes place in a duct such as a rocket combustion chamber it is necessary to take into account that the overall velocity of the gas stream changes as the liquid fuel is converted to the burned gases. In addition the duct may be of variable area. As a result it is necessary to incorporate an additional equation so that the variation of velocity, or elapsed time, can be estimated. This is the overall continuity equation, which is essentially a mass balance for all material flowing across a unit area, thus

$$\rho_g V_g + \rho_L V_L = M = \text{constant} \qquad (5.10)$$

67

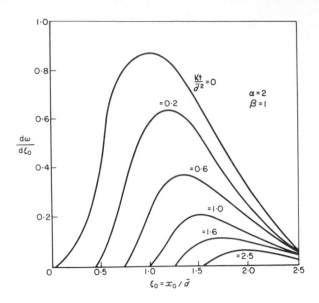

Figure 5.1. *The change of weight distribution with duration of combustion for the case where α = 2, β = 1 (after Tanasawa and Tesima, 1958).*

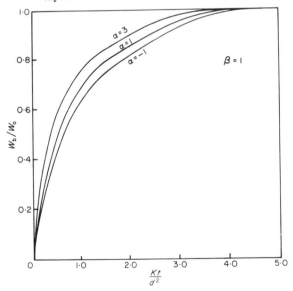

Figure 5.2. *The change of the extent of combustion with the duration of combustion. Combustion is complete when $W_b/W_o \simeq 1$ (after Tanasawa and Tesima, 1958).*

where M is the total mass flow / unit area. Any decrease in the liquid phase mass flow rate $(\rho_L V_L)$ is compensated exactly by an increase in the gas mass flow rate $(\rho_g V_g)$. Therefore

$$\rho_g V_g = M (1 - r^3/r_o^3) \qquad (5.11)$$

Hence, the gas velocity is connected with the local droplet size, r, by this simple relation.

Spalding (1959) used this approach together with the 'd^2 droplet burning law' (Equation 4.3) and the Stokes drag law (Equation 2.20) and applied it to the analysis of the performance of a liquid-fuelled rocket engine combustion chamber. In a rocket combustion chamber the liquid fuel, such as kerosine, is injected into the chamber from the end wall as shown in Figure 5.3 and the length of the combustion chamber must be

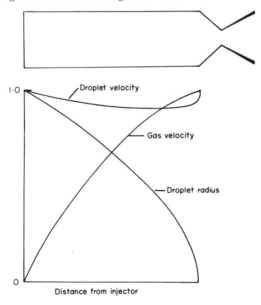

Figure 5.3. *Variation of droplet size, droplet velocity and gas velocity along a rocket combustion chamber (after Spalding, 1958).*

such that droplet burn-out occurs immediately prior to the expansion part of the chamber. If the chamber is too short then loss of performance results because of incomplete combustion; if it is too long then the rocket chamber is unnecessarily heavy. Spalding undertook an

69

analysis based on the assumption that the droplets are mono-sized and considered the effects of droplet size, droplet injection velocity and properties of the fuel and oxidant. Solutions were obtained on the variation of droplet velocity, gas velocity and droplet velocity as a function of distance from the injector. A typical solution for propellant vaporisation is shown in Figure 5.3. It was shown that (a) the combustion chamber length necessary for complete vaporisation is proportional to the square of the initial droplet radius, (b) increasing the injection velocity (the forward injection velocity from the end plate) will increase the length required for complete evaporation, and (c) preheating the propellants is of advantage in that it assists the rate of combustion.

A number of more sophisticated models of spray combustion have been developed in conjunction with various national space rocket programmes which take into account injection, atomisation, complex drop size distribution and realistic droplet ballistics including droplet-droplet collisions. Such analyses are too complex to permit analytical solutions and computer solutions are necessary.

5.3   COMBUSTION IN LAMINAR FLAMES OF SPRAYS AND MISTS

A mixture of spray and air is able to support a propagating laminar flame in the same way as can a premixed gaseous mixture. The simplest picture is presented by the combustion of a spray mixture in a tube in which the flame travels down the tube with a laminar, planar flame front at right angles to the direction of flow. This flame travels at a constant velocity characteristic of the mixture and its temperature; this is the laminar burning velocity. Likewise, stationary flames can be stabilised on a burner and, like premixed flames, the flames may undergo laminar or turbulent combustion.

Two modes of combustion have been identified in laminar flames. If the spray is a lean mixture and consists of small droplets of volatile fuels (such as gas turbine fuels), roughly less than 10μm in diameter, then the droplets vaporise and mix with the air before the flame front is reached and behave as premixed flames. Such flames differ little from premixed flames and have been termed 'homogeneous flames'.

When the droplets become too large for this to occur they reach the flame front with the evaporation and mixing process incomplete, and in the case of large involatile droplets (such as heavy fuel oils) evaporation and mixing may be negligible. This results in the collapse

of the coherent laminar flame front and the establishment of a
combustion system based on droplet combustion. This form of combustion
approximates to the case given in Figure 1.1 and such flames have been
termed brush or heterogeneous flames.

A number of theories of laminar spray flames have been developed
with the objective of deriving a mathematical solution for the
heterogeneous burning velocity. The procedure involves setting up
differential equations for the rate at which the spray burns throughout
the combustion zone, the equation being based on the d^2 law and the
droplet distribution law. In the case of mono-sized droplets this
equation would take the form

$$\frac{dZ}{dx} = - \int_0^\infty \rho_L 4\pi r \qquad (5.12)$$

where x is the distance through the flame and Z the mass fraction in the
gas phase. When $Z = 1$ droplet combustion is complete. Likewise an
expression can be derived to give the increase of temperature throughout
the flame; this equation is based on the extent of droplet combustion
(i.e. Equation 5.12) and the heat of combustion. From solutions of
these equations, and using the assumption that droplet ignition occurs
when a critical droplet temperature is reached, then it can be shown
that the mass burning rate is given by

$$M \propto \frac{1}{r_0} \frac{1}{\rho_L^{\frac{1}{2}}} \frac{\lambda}{C_P} \sqrt{(P W_0)} \qquad (5.13)$$

where γ and C_P refer to the average gas phase values of the thermal
conductivity and specific heat respectively, P is the pressure and W_0
the initial mean molecular weight.

The variation of M with r_0 implies that the burning velocity
increases with decrease in droplet size provided that the same quantity
of fuel is present. Obviously, the equation is no longer applicable
once the droplet diameter falls below a certain critical size, since the
combustion mechanism changes to the homogeneous mode.

An alternative approach to the prediction of burning velocities is
based on the appreciation of the network (or 'relay') theory of droplet
ignition. In this mechanism the flame propagates discontinuously from
one droplet to another. An analysis by Mitzutani and Ogasawara (1965)
is based on a one-dimensional flame model as shown in Figure 5.4. Here
all the droplets in one plane, A, are assumed to ignite at the same
instant and the flames surrounding each droplet are replaced by a

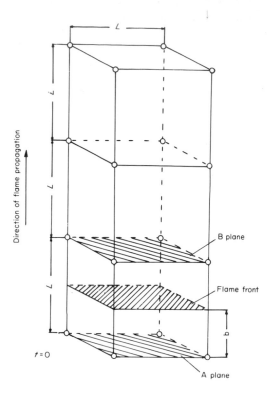

Figure 5.4. *Network model of flame propagation (after Mizutani and Ogasawara, 1965).*

hypothetical planar flame front. Ignition of the adjacent B plane of droplets then occurs after a time lag corresponding to the time required for these droplets to heat up and to ignite, that is after the elapse of the ignition delay time. In this way the flame propagation velocity may be derived from this time interval and the known number concentration of droplets in the flame. This latter quantity, n, is simply derived for mono-sized droplets from the relationship

$$n = \frac{\text{mass of spray unit volume}}{\text{mass of one droplet}} = \frac{3C_s}{4\,r_L{}^3\rho_L} \qquad (5.14)$$

In this way it has been possible to derive flame propagation velocities as functions of droplet size. Experimentally it is known that flame propagation velocities become smaller with increase in droplet diameter and that it rapidly increases as the diameter increases more or less linearly along with the mixture concentration for a fixed diameter. These effects are indicated in Figure 5.5 and the 'network' theory is capable of predicting these general trends.

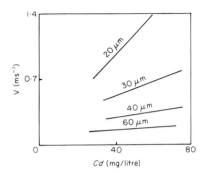

Figure 5.5.   Variation of burning velocity, V, with mass concentration, Cd, and initial droplet size, d_o.

5.4   LIMITS OF INFLAMMABILITY

The network theory already is also applicable to the prediction of the lower inflammability limits of sprays.   On this basis, in order for stable flame propagation to occur it is necessary for the droplets in plane B (see Figure 5.4) to be ignited before the droplets in plane A have been burnt out.   Inflammability limits in sprays are consequently a function of the mass concentration of the spray, the droplet diameter and the direction of flame propagation.   The latter effect is quite marked in sprays and varies with droplet sizes since it is largely caused by the sedimentation of the droplets.   The variation of the lower inflammability limit as a function of droplet size is shown in Figure 5.6, it should be noted that as the droplet size becomes smaller

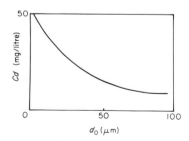

Figure 5.6.   Variation of lower inflammability limit expressed as a mass concentration, Cd, with droplet size, d_o.

the behaviour tends to that of a premixed gaseous mixture. In the case of the rich inflammability limits, the conditions are more complex since the droplets make up a substantial proportion of the combustible mixture.   In general, for heterogeneous combustion, rich spray mixtures can be burned which are richer than the equivalent premixed mixture. This arises because the spray flame as a whole is supported by diffusion flames surrounding individual droplets, so that the bulk of the liquid in the spray does not become involved in combustion until it is evaporated from the droplet surfaces.   It is, however, taken into account in the calculation of the overall stoichiometry (or air/fuel ratio).   Thus the flame passes through the spray via the network process leaving cells of unburned mixture.   The behaviour of sprays under rich limit conditions has not yet been subjected to theoretical analysis.

The rich limit behaviour of small droplets burning under homogeneous conditions is again similar to the equivalent premixed gaseous flame behaviour.

6 Applications of Spray Combustion

Oil burners are used in a wide range of industrial plant, such as furnaces and boilers, for the heating of materials either directly or indirectly. Examples of direct heating are in kilns and furnaces generally, whilst the most common applications of indirect heating are in steam raising and process heating within the chemical industry. In all these applications, liquid fuels have a number of attractive features, the major ones being the ease of handling and storage of the fuel and the highly radiant nature of the flames produced which can be of great advantage in certain applications.

Spray combustion is also widely used for transportation purposes, finding applications in compression ignition (diesel) engines and in aviation, marine and industrial gas turbines. Again the ease of fuel handling and storage makes it an attractive fuel for such applications.

6.1 GENERAL FEATURES OF INDUSTRIAL OIL BURNERS

The essential components of an industrial burner are:

(a) an atomiser to produce the spray,

(b) an air register which admits the air and determines the flow pattern to promote good mixing,

(c) the burner throat and a stabiliser to ensure stable combustion.

Burner units having oil throughputs in the range of 10-350 kg h^{-1} or heat inputs of 300 kW to 6 MW (0.5 to 20×10^6 Btu h^{-1}) are available as packaged burners containing the burner and the fuel and air supplies as well as the ignition device and all the controls. Such a unit is shown in Plate 6.1. Smaller units, such as central heating units, and larger units, such as in very large steam generating plant, are usually available as specialist units.

The mixing of the fuel and the air depends on the way in which the oil is atomised, on the flow pattern of the combustion air supply which is determined by the design of the register and on recirculation. These detailed factors are discussed in Section 6.2 and the more general aspects of spray combustion are considered in the present section.

Plate 6.1.   *A packaged burner.   A Weishaupt type GL7 dual-fuel burner (butane/35 sec oil) firing a boiler rated at 1575 kg steam h^{-1}.*

6.1.1   Combustion Stoichiometries and Flame Temperatures

If the elemental composition of a fuel oil is known it is possible to calculate the overall stoichiometry of the combustion process.   Thus a fuel oil containing, for example, 84% by weight of carbon, 12% by weight of hydrogen, 1.5% by weight of oxygen and 2.5% by weight of sulphur would undergo the following overall reaction.

$$10.031 \; O_2 + 37.7 \; N_2 + C_7 \; H_{12} \; O_{0.0938} \; S_{0.0781} \quad =$$

$$7 \; CO_2 + 6 \; H_2O + 0.0781 \; SO_2 + 37.7 \; N_2$$

Generally sprays are burned with an excess amount of air termed 'excess air', so that smoke formation is minimised and this air would be additional to the quantity given above.

At flame temperatures the combustion products are dissociated, principally by the dissociation reactions

$$CO_2 \overset{1}{\underset{\rightleftharpoons}{}} CO + \tfrac{1}{2} O_2$$

$$H_2O \overset{2}{\underset{\rightleftharpoons}{}} H_2 + \tfrac{1}{2} O_2$$

$$H_2O \overset{3}{\underset{\rightleftharpoons}{}} \tfrac{1}{2} H_2 + OH$$

$$\tfrac{1}{2} H_2 \overset{4}{\underset{\rightleftharpoons}{}} H$$

$$\tfrac{1}{2} O_2 \overset{5}{\underset{\rightleftharpoons}{}} O$$

together with

$$\tfrac{1}{2} N_2 + O_2 \overset{6}{\underset{\rightleftharpoons}{}} NO$$

Consequently, the products at the flame temperatures contain small quantities of carbon monoxide, hydrogen etc. and the amounts of these products may be estimated from standard tables (Spiers, 1962) or by computation by means of standard techniques (see for example, Gaydon and Wolfhard, 1970) for which small programmable calculators are eminently suitable.

Theoretical adiabatic flame temperatures may also be derived from the heat released by the combustion process. The enthalpy change associated with the combustion process must take into account dissociation. Calculations of this type have shown that the maximum theoretical flame temperatures for the combustion of a n-heptane spray with air are 2100 K for zero excess air stoichiometric mixture, 1900 K for 10% excess air and 1800 K for 20% excess air. Heavy fuel oils would have a slightly lower flame temperature which is dependent upon the ash and sulphur contents. In practical flames, heat losses to the furnace and in the waste gases greatly reduce the flame temperatures actually attained. These temperatures can be obtained only by experimental techniques or by heat balance calculations; details of these techniques are given in standard text books such as Thring (1962). Generally flame temperatures are in the region 1300 K to 1900 K.

6.1.2 Combustion Intensities

The combustion intensity, for any combustion system is given by Equation 6.1

$$I = \frac{F \cdot H}{V_c \, P} \quad \text{Wm}^{-3} \text{s}^{-1} \text{bar}^{-1} \tag{6.1}$$

where F is the fuel feed or firing rate, H is the enthalpy of combustion, V_c the chamber volume and P the pressure. Obviously the smaller the combustion volume for a given fuel feed rate the higher the combustion intensity. In spray combustion, the rate of energy release is virtually totally controlled by droplet burning, any burn-out of residual carbon monoxide first formed by droplet combustion makes only a small contribution to the total heat release. On this basis the combustion intensity may be calculated as follows. If the air quantity is Q then at stoichiometric firing the air rate is FQ, thus with $E\%$ excess air the total rate is $FQ \, (1 + E/100)$. If there is little change in volume due to the combustion products, then the hot volume is given by

$$\frac{FQ \ (1 + E/100)}{P}\left(\frac{T_f}{T_o}\right) \quad \text{m}^3 \text{ s}^{-1} \text{ bar}^{-1} \tag{6.2}$$

where T_f and T_o are the flame and ititial temperatures respectively. Thus

$$V_c = \frac{FQ}{P}(1 + E/100)\left(\frac{T_f}{T_o}\right) t_B \tag{6.3}$$

and

$$I = \frac{H}{Q \ (1 + E/100)(T_f T_b) t_B} \quad \text{Wm}^{-3} \text{s}^{-1} \text{ bar}^{-1} \tag{6.4}$$

Generally H/Q is approximately constant so the combustion intensity depends primarily on t_B which is given by Equation 5.2. For droplet diameters in the region of 100 μm in diameter and assuming a reasonable value for K (say 0.5 mm^{2}s^{-1}) and plug flow, then combustion intensities are in the region of 10^5kWm^{-3}bar^{-1} for the combustion of volatile liquids and one or two orders of magnitude less for heavier fuels.

6.2   AERODYNAMIC FEATURES OF SPRAY FLAMES IN STATIONARY COMBUSTION UNITS

Whilst many useful deductions relating to overall effects such as efficiency and combustion intensity may be made on the basis of the assumption of plug flow, for many aspects it is necessary to have a more detailed knowledge of the aerodynamic interaction of the fuel spray and the air stream.

Industrial oil flames may be broadly classified into two groups depending upon the method of atomisation used:

(a) Turbulent jet diffusion flames in which the oil is atomised in a twin-fluid atomiser so that the primary air-fuel spray has a sufficiently high momentum to dominate the combustion process.

(b) Pressure jet and rotary cup atomisers in which the momentum of the fuel spray is low compared with the momentum of the air flow.

These two categories are discussed in more detail in standard texts on aerodynamics (e.g. Beér and Chigier, 1973) and the essential features are outlined further below.

6.2.1   Twin Fluid Atomisers

The flame produced in burners without swirl is a long narrow angle turbulent jet diffusion flame of the type shown in Figure 6.1 and the spread of the flame is largely due to turbulent diffusion of the atomising air.   Since the droplet diameters are generally small

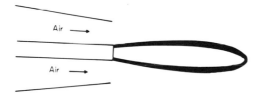

Figure 6.1.   *A typical flame produced by a twin fluid atomiser.   The flame would generally be turbulent.*

($d_{32} \simeq$ 40-50 μm) the flame behaviour is controlled by the overall jet momentum flux but of which the largest component is that of the primary atomising air stream, the droplets, of course, largely following the air flow.   This controls the local fuel/air mixture, which in turn controls the amount of soot produced and hence the flame emissivity.

Consequently, burner designs are frequently based on some method which optimises the jet momentum with the minimum expenditure of energy for atomisation.

The flame shape can be deduced by considering the behaviour of an equivalent jet of air and Thring (1962) has shown that the flame length (ls) is given by

$$ls \simeq 5.3 \ d_e/C_s \qquad (6.5)$$

where C_s is the relative concentration of the fuel in the stoichiometric mixture and d_e is the diameter of a nozzle through which the fuel and air would flow if their density were that of the flame gases which is given by

$$d_e = \frac{2(m_o + m_s)}{(\pi \rho_f G)^{\frac{1}{2}}} \qquad (6.6)$$

where m_o and m_s are the masses of the air and spray, ρ_f the density of the flame gases and G the total momentum.

If the air stream is swirled, then the degree of swirl determines the flame spread. High degrees of swirl result in short but wider angled flames. As the degree of swirl increases, recirculation along the central axis occurs in which the hot combustion products flow back towards the atomiser, resulting in the rapid heating up of the combustible mixture. The behaviour of swirling gas streams is outlined in the book by Beer and Chigier, 1973.

6.2.2   Pressure Jet Burners

In a pressure jet burner, the atomiser is surrounded by an air stream directed by a register device as illustrated in Figure 6.2. Since the momentum of the liquid spray is relatively low the aerodynamic behaviour is a complex function of the air flow pattern. Because of the large number of arrangements possible numerous studies have had to be undertaken to obtain some understanding of these systems. The principal investigations have been made by the International Flame Research Foundation at Ijmunden and numerous reports have been published which have been summarised by Beer and Chigier (1973). The factors controlling the shape of the flame and its stability are the extent of swirl, the type of flame stabiliser required to anchor the flame and also the combustion chamber shape.

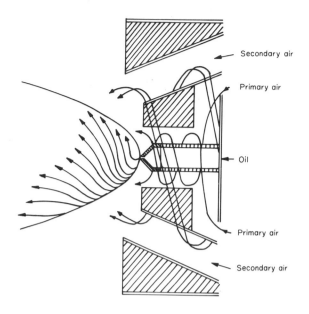

Secondary air

Primary air

Oil

Primary air

Secondary air

Figure 6.2. *Formation of a flame with swirl.   The arrangement*
shows a pressure jet atomiser in a Hamworthy HPD
register which contains two separate concentric
swirling flows.

The influence of a significant amount of air swirl is to cause
internal recirculation as indicated in Figure 6.3.   The effects of
various flame stabiliser devices are shown in Figure 6.4.   It is of
interest to note that the recirculation of combustion gases can play a
significant role in the mechanism of the atomisation process
(Dombrowski *et al.*, 1974) and data obtained for atomisers under cold
conditions may not necessarily apply to combustion chambers.

An important aspect controlling the combustion of pressure jet
flames, particularly with regard to the formation of smoke and the
radiative properties of the flame, is the way in which the air is
entrained into the flame.   Some typical pressure jet flames are shown
in Plates 6.2 and 6.3, and the luminous nature of the flame is apparent.
Numerous studies have been made of the air entrainment into such flames,
the typical structure of a dense flame with only limited air entrainment
is shown in Figure 6.5   It is clear that in this case the classical
concept of a droplet-burning controlled spray flame is not applicable
to the whole of the flame.   In a considerable number of applications

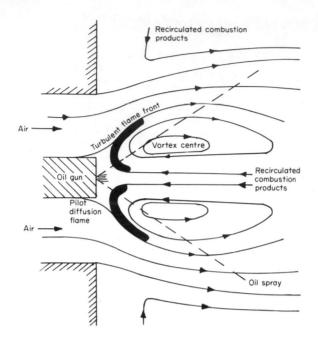

Figure 6.3.   *Schematic representation of internal recirculation and its influence on the position of the flame front (after Beér and Chigier).*

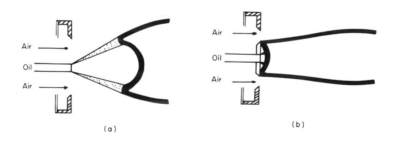

(a)

(b)

Figure 6.4.   *Stabilisation of pressure jet flames in (a) a simple pressure jet system, and (b) by a stabiliser plate or baffle. The flame shapes in both cases are determined to a large extent by the relative momenta of the air and liquid flows.*

Plate 6.2. *A spray flame produced by a pressure jet burner.*

a number of pressure jet burners are used which fire from the same
wall. Frequently these are mounted only several burner diameters
apart and there may be interaction between the flames. Generally
there is a shortening of the flames due to enhanced heat transfer and
the interaction of the droplets with reversed flow.

6.3 <u>HIGH INTENSITY COMBUSTORS</u>

With conventional pressure jet burners the flame obtained is
always somewhat ragged and waving and is highly luminous. The heat
release therefore is such that the heat flux from the flame is
non-uniform. Consequently, a number of burner systems have been
developed utilising a high degree of recirculation, frequently
involving a double-vortex flow which gives a very stable flame with
very high combustion intensities. A typical arrangement is shown in
Figure 6.6. Essentially the combustion takes place in a precombustion
chamber and recirculation within it is caused by the high entry

*Plate 6.3.   A spray flame produced by a swirl pressure jet burner.
This is an end-on view and the hollow nature of the
flame is apparent.*

velocity of the air.   The combustion gases are then ejected into the
furnace oxygen with a high velocity and this assists convective heat
transfer.   The high levels of recirculation and good resultant mixing
ensure low soot concentrations in the burned gases.

6.4   FURNACE APPLICATIONS

Oil burners are widely used in a wide variety of industrial
furnaces, principally steel furnaces, glass tanks and a multitude of
smaller applications such as cement or pottery kilns, enamelling and
reheating furnaces etc.   Generally twin-fluid atomisers are used for
furnace applications with pressure jet atomisers being restricted to
on-off applications because of their limited turn-down ratio.   The air
blast atomiser offers great flexibility because of the amount of preheat
that can be supplied, this being achieved by the use of recuperators.
In all applications of preheat it is very important to balance the
amount of preheat with the amount of excess air used, since any
advantages gained in efficiency from preheat may be lost if excessive
excess air is used.

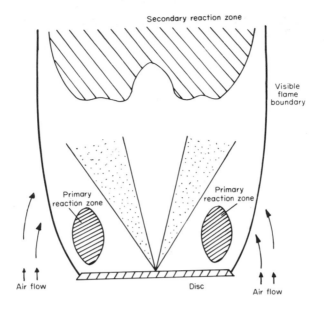

Figure 6.5.   *Structure of a dense pressure jet flame (after McCreath and Chigier, 1973).*

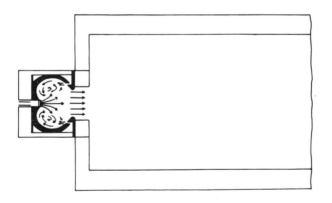

Figure 6.6.   *A typical toroidal chamber arrangement.*

85

The burners used may involve fully automatic, semi-automatic or manual controls depending upon the application. Many processes require accurate control of the furnace temperature. Generally this is achieved by the controlled use of excess air or by the recirculation of flue gases, the latter being the preferred technique since it has a higher efficiency because there is no increased heat loss in the flue gases. Close control of the fuel/air ratio is also essential in any furnace in order to obtain maximum efficiency. Generally an oxygen or CO_2 analyser is used to monitor the flue or furnace gas composition. Furthermore, it may be used to adjust the fuel/air ratio control whenever necessary.

The type of burner and atomiser vary according to the application. Of major importance is the heat flux characteristic curve. In conventional spray flames the peak heat transfer occurs in the zone of most intense combustion at the beginning of the flame as shown in Figure 6.7. As droplet and soot burn-out proceeds the heat flux decreases as indicated. In toroidal type combustion chambers the heat flux curve is more uniform.

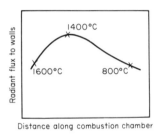

Figure 6.7. *Variation of the radiant flux to the walls of a combustion chamber as a function of distance along the chamber. Typical gas temperatures are also indicated.*

Typical furnace applications include the following:

(a) Metallurgical Industries

The spray combustion of liquid fuels, particularly involving air blast atomisers, is widely used in the iron and steel industries for applications such as billet reheating furnaces, direct fired heat treatment furnaces, hardening and tempering furnaces, forge furnaces, soaking pits, blast furnaces and a wide range of other metallurgical applications, many of which require carefully controlled atmospheres of combustion products.

86

Non-ferrous applications include reverbatory furnaces and melting furnaces in general. Plate 6.4 illustrates a typical furnace application. Oxy-oil burners are used in a variety of applications in the steel and non-ferrous metal industries for the melting of scrap and processing. A typical high intensity flame produced by a Shell toroidal burner is shown in Plate 6.5.

(b) Ceramic and Glass Industries

Air blast atomisers are frequently used in the heating of glass tanks, brick and tunnel kilns. In cement kilns, pressure jet atomisers having capacities of several tons per hour may be used.

(c) General Manufacturing Industries

Spray combustion is widely used for air driers for drying grain, sand and foodstuffs, for evaporating brine, for paint drying ovens, for rotary calcining and a wide variety of other applications.

6.5   BOILER APPLICATIONS AND PROCESS HEATING

6.5.1   Steam Raising

The major application here is for large water tube boilers in thermal power stations but there are countless applications in smaller steam raising plant and central heating complexes.

The boilers are specially designed to match the heat transfer characteristics of the flames used and employ a first section designed for radiant heat transfer and a convective heat transfer second stage.

In thermal power stations the combustion chamber is essentially rectangular in shape and a number of burners may fire from one wall, or from opposing walls or even from the corners. A typical combustion chamber shape and standard oil burner are shown in Figures 6.8 and 6.9. The oil used in these installations is a residual fuel of viscosity 350 - 900 cSt at 37.8°C (100°F) and having a sulphur content of about 2-3%, although in very special applications a gas oil may be used. Handling of the fuel therefore involves heated tanks with steam traced or electrically heated pipework. The type of atomiser used varies widely but is usually a pressure jet or assisted pressure jet atomiser rather than a twin-fluid atomiser as such.

Spray combustion is also widely used in boilers varying from small packaged units to large installations. These may involve Shell boilers of the Lancashire or Economic type or large water tube boilers with ratings from 500 to 20 000 kg/h are often used for oil combustion. At one time, pressure jet atomisers were widely used but rotary cup burners are extremely well suited for boilers as are twin fluid

Plate 6.4.   Oil burners being used in furnace heating.

Plate 6.5.   A flame produced by a Shell toroidal oxy-fuel burner.

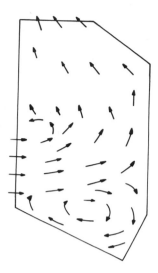

*Figure 6.8. Typical power station combustion shape and basic
flow pattern. In such a unit 30-40 large burners would
fire into a chamber having a volume of 5000 m³ or more.*

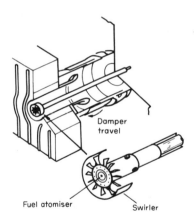

*Figure 6.9. Typical oil gun used in a power station
(after Anson and Tindall, 1967).*

atomisers.   The fuels used may range from 200 cSt to 900 cSt at 37.8°C and are heated prior to injection to the boiler.

6.5.2   Process Heating

Oil burners are used in a variety of applications in the petroleum and chemical industries where liquids are heated.   Typical examples here are pipe still heaters, reforming units etc. and other fired heaters.   Other applications include packaged air heaters for the drying of foods, submerged combustion units for the evaporation of liquids and also fuel oil gasification applications for the preparation of controlled or inert atmospheres.

An example of toroidal combustion chambers used in a refinery is shown in Plate 6.6.

6.6   COMPRESSION IGNITION ENGINES

Another major application of spray combustion is the compression ignition engine.   Here the fuel spray is injected into air which is heated by the compression stroke of the piston;   its ignition is controlled by the chemical composition of the fuel and thus the fuel properties play a key role.   The processes of ignition and combustion that occur will be described below with reference to the four stroke engine cycle.

The basic four stroke diesel cycle involves the following processes, commencing with the piston at the top of its upward travel having completed an exhaust stroke:

(a) Air is drawn into the cylinder on the downward stroke.

(b) During the next stroke the air charge is compressed to a final pressure of 35-50 bar with a final temperature of about 900-1000 K;   in turbocharged engines the pressures and temperatures are higher depending upon the degree of turbocharging.   Fuel is injected by means of a pressure jet atomiser in the form of a fine spray shortly before the piston reaches the end of the compression stroke.   After the elapse of the ignition delay time ignition occurs and combustion results at about the end of the compression stroke.   Pressures may exceed 70 bar and temperatures may exceed 2200 K.

(c) Energy release continues as work is done during the downward expansion stroke, the amount of fuel injected having been metered according to the power required, that is, the throttle setting.

(d) The exhaust valve opens at the end of the piston travel on the expansion stroke and the burned gases are expelled from the cylinder.

Plate 6.6.   A 190 MW (650 million Btu h $^{-1}$) crude oil heating furnace at the BP Europort refinery.   The furnace (by Birwelco) is fired by 10 Urquhart dual-fuel single toroidal combustors.

One important design feature of the combustion chamber is of considerable importance to the combustion process.   With 'direct injection' (D.I.) the fuel is injected directly into the main combustion chamber as shown in Figure 6.10.   With indirect injection (I.D.I.) the fuel charge is directed into a small precombustion chamber where the fuel spray first ignites and spreads to the main chamber where additional air is available to sustain combustion.   A variety of designs are available but the general form of indirect injection is also shown in Figure 6.10.

In direct injection diesel engines, fuel is injected into swirling air, the degree of swirl being a function of the engine design. The atomiser used may be a simple single hole pressure jet or, in the

(a)

(b)

Figure 6.10. *Basic forms of compression ignition combustion chambers. (a) Direct injection, (b) indirect injection. Many other configurations are possible.*

case of large engines, it may contain a number of holes, these holes being angled so as to distribute the sprays produced throughout the combustion chamber.

6.6.1 The Ignition Process

When the spray is injected into the hot air in the cylinder the ignition processes already outlined in Section 4.6 take place. After the elapse of the ignition delay period, as indicated in Figure 6.11(a), ignition occurs and the pressure in the combustion chamber rises as further fuel is injected until the end of the injection period. If the ignition delay period is too long then too much fuel is injected prior to ignition and 'knock' may occur. The ignition quality of a diesel fuel is characterised by the cetane number (Hobson and Pohl, 1974) which is related to the ignition delay time. Fuels with a short delay time, such as a reactive alkane of which n-hexadecane (cetane) is the standard fuel, have a high 'cetane number'; unreactive fuels such as stable aromatic molecules of which α-methyl naphthalene is the standard fuel, have a low 'cetane number'.

Once ignition occurs the fuel that is subsequently injected is ignited by the network mechanism and this is largely independent of the fuel type. The fuel droplets then continue to burn by the 'd^2 law mechanism', a process again largely insensitive to the composition of the fuel. The heat release rate curve shown in Figure 6.11(b) which

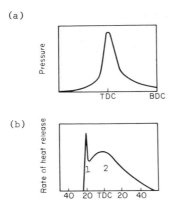

(a)

(b)

Figure 6.11. *Representation of (a) a P-t indicator diagram*
for a compression ignition engine and also
(b) the heat release curve.

may be determined from the experimentally derived pressure-time
indicator curve, shows the occurrence of the various stages.
Photographic studies have shown that ignition occurs in the outer
periphery of the injected spray, or at the leading edge of a swirled
spray and then proceeds through the remainder of the combustible
mixture.

The initial ignition and rapid heat release rate shown in region 1
in Figure 6.11(b) is associated with the combustion of the evaporated
fuel, and also the smaller droplets, by the homogeneous mechanism.
This region burns as in premixed gaseous combustion with a flame of low
luminosity. At the point of maximum heat release, the cumulative heat
release is about 5% of the total heat release, indicating the overall
amount of fuel in vapour form (or small droplets) prior to ignition.

The next part of the heat release curve is associated with the
normal 'd^2 law' burning of the droplets in region 2. The final part of
the curve is associated with the burning of soot particles produced
during heterogeneous combustion and any carbon monoxide produced. The
combustion occurring throughout these two regions is associated with
high flame luminosity.

A number of theoretical models have been developed to predict the
rate of combustion of sprays injected into a diesel combustion chamber
(Khan, 1972; Crookes *et al.*, 1973). These models are based on models
predicting the rate of evaporation of the liquid fuel spray and the rate
of mixing of the spray with air. The latter quantity is obtained on

93

the basis of applying the conservation of energy momentum equations together with the rate of spread of the jet. This type of approach estimates the local fuel-air ratio thus enabling the rate of heat release and rate of pollutant formation to be estimated.

6.6.2   Spray Combustion in Swirling Air

In the direct injection compression ignition engine, fuel is sprayed from a multihole nozzle into a toroidal bowl in the piston head. In the larger, slower engines the hot compressed air in the piston head bowl is virtually quiescent and mixing is by air entrainment into the spray. In most automotive engines which have higher speeds and therefore low combustion chamber residence times, the time for combustion is short. Thus to speed up mixture formation and droplet burning the relative velocity between the droplets and the charge air is increased by introducing a high air swirl velocity component in the bowl. This is achieved by the piston motion and the inlet point design. Since the spray consists of a large number of different sized droplets the relatively large droplets are concentrated in the core and trailing edge of the swirled spray. The very small droplets and vapour are carried away with the air and form the leading edge of the spray. Thus the local fuel-air ratio varies within a swirled spray varying from zero at the leading edge to a maximum in the core of the spray. Ignition of the mixture occurs just after the leading edge where the fuel vapour-air ratio is within the inflammability limit and there is sufficient contact with the hot air, and the combustion proceeds through the remainder of the combustible mixture.

6.7   GAS TURBINES

The conventional gas turbine combustion chamber is required to burn the fuel with air supplied at high velocities from a compressor and generally combustion must be undertaken in as small a volume as possible, that is, high intensity combustion coupled with high flame stability is required. This has been achieved in practice by means of the arrangement shown in Figure 6.12. In some engines the combustion chamber takes the form shown in Figure 6.13 and in that case there are a number of such combustion chambers disposed around the engine as shown in Figure 6.14(a). This is called the tubo-annular combustion chamber. Each combustion chamber has an inner flame tube around which there is an air casing and in the tubo-annular system the flame tubes are all interconnected. This allows each tube to operate at the same pressure and permits the flame to propagate around the flame tubes during ignition.

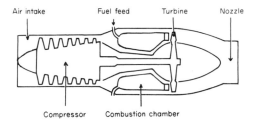

Figure 6.12. *Outline of the essential features of a gas turbine.*

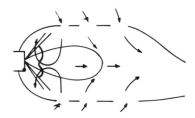

Figure 6.13. *Flow patterns in a gas turbine combustion chamber. The region of smoke formation and primary combustion are also indicated.*

Some engines have an annular combustion chamber in which the flame tube takes the form of an annulus around the engine, open at one end to the compressor and at the other end to the turbine nozzles as indicated in Figure 6.14(b). An inner and outer air casing contain the flame tube and holes in the flame tube walls allow cooling air to enter the tube.

The combustion chamber must be able to burn efficiently over a wide range of operating conditions and combustion must be stable so that flame blow-out does not occur. In addition, the burned gases leaving the combustion chamber must be suitably cooled by dilution with air so that the turbine blades are not destroyed.

The way in which this is achieved in the combustion chamber is as follows. Air from the engine compressor enters the combustion chamber

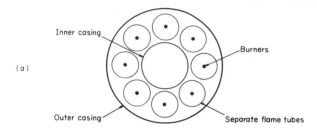

(a)

Inner casing

Burners

Outer casing

Separate flame tubes

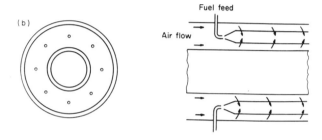

(b)

Fuel feed

Air flow

Figure 6.14.   *Arrangements of combustion chambers in (a) a*
tubo-annular system, and (b) an annular
combustion chamber.

at a velocity up to 150 m s^{-1} and this is decelerated by the diffuser
to about 30 m s^{-1}.   Even so since this velocity is much greater than
the burning velocity of the kerosine spray, which is of the order of
0.4 m s^{-1}, the flame has to be stabilised by a high degree of
recirculation in the primary zone as indicated in Figure 6.12.
Typically 20 - 30% of the total air flow is introduced into the
combustor in this zone and the kerosine is injected as a spray so that
combustion occurs with a near-stoichiometric mixture in this primary
zone.   Since the stoichiometric air-fuel (mass) ratio is 15:1 and the
overall air-fuel ratio is required to vary between 45:1 and 130:1 the
additional air is added in the secondary and dilution zones as shown in
Figure 6.13.

The fuel spray from the atomiser (termed an injector) is injected
so that it intersects the recirculation vortex at its centre.   The

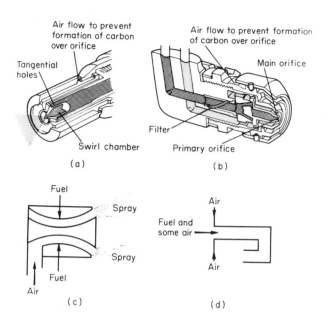

Figure 6.15. *Gas turbine atomiser types.* *(a) Simplex,*
(b) Duple, (c) air spray, and (d) vaporiser tube.

most common form of fuel injection is by means of a pressure atomiser,
although vaporisers are also used and more recently air-blast burners
have been adopted.

Pressure jet atomisers may be of the Simplex or Duple types which
are illustrated in Figure 6.15, both producing a hollow core spray with
mean droplet sizes of about 80 μm. The former type has a limited turndown ratio whilst the latter uses a main and a pilot (or primary) spray
to obtain a wider operating range. The main difficulty with these types
of burners is that many of the present-day engines, which operate at high
combustion chamber pressures, would result in the formation of soot
because of the rich mixtures formed in the central part of the spray as
shown in Figure 6.13. More recently air-blast atomisers of the type
shown in Figure 6.15 have been adopted in some engines such as the
Rolls-Royce RB 211. These are essentially air-blast atomisers and
incorporate a pintle and a prefilming surface, the resultant spray is
thus not so fuel-rich in the central regions of the spray and, furthermore,
the droplets are smaller (≈ 40 μm), both factors contributing to the
reduction of smoke formation.

The third type of injector is the vaporiser system illustrated in Figure 6.15. These systems do not in fact vaporise all the fuel, except at low fuel flows and at maximum flow only some 10% of the fuel is evaporated. Usually a small quantity of air passes through the vaporiser to prevent carbon build-up and essentially it behaves as an air blast atomiser. However, large droplet sizes result and they are not suitable for high pressure engines. More recently a premix atomiser (or carburettor) has been developed in which a considerable amount of air is passed through a spiral vaporiser and the fuel/air mixture injected tangentially to the incoming air stream. Another application of spray combustion in aircraft gas turbines is in 'afterburning' or 'reheat' which is a method of augmenting the basic thrust of an engine to improve take-off, climb etc. Reheat consists of the burning of an injected spray between the engine turbine and the jet pipe and using the unburned oxygen in the exhaust gases to support combustion. The resultant increase in gas temperature and consequential increased velocity in the jet leaving the propelling nozzle increases the engine thrust.

An atomised fuel spray is fed into the jet pipe through a number of burners which are arranged to distribute the fuel evenly over the flame area. Typically the fan spray type of atomiser may be used and generally it is mounted on a circular pipe mounted centrally in the jet pipe. Because of the high gas velocity entering the jet pipe, which may be of the order of $200 - 400$ m s^{-1}, it is not possible to obtain a stable flame without some form of flame-holding device. Generally a circular V-shaped gutter is used for flame retention, as indicated diagrammatically in Figure 6.16, in which the local gas velocity is sufficient to permit combustion.

Ignition of the injected spray has to be achieved by means of an electrical spark or catalytic device because although the gas temperatures present ($c.$, 700°C) are sufficient to cause ignition, the ignition delay is so long that the gases would have been exhausted from the jet pipe before ignition had actually occurred.

Reheat is often accompanied by an instability termed 'buzz'. Buzz is a low frequency instability which can result in the flame being extinguished. It apparently arises from a finite-amplitude pressure perturbation of the mixing and combustion processes so that the fuel-air mixture is incompletely burned. This unburned mixture then passes through the duct causing a new pressure pulse, resulting in the maintenance of the oscillation. Its origins, however, may lie in the

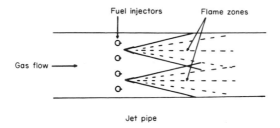

Figure 6.16. *Method of fuel injection and flame stabilisation during reheat.*

transient nature of the fuel injection process. This type of instability also occurs in liquid-fuelled rocket motors and is generally termed 'entropy instability' because the fluid motion must carry an entropy inhomogeneity.

6.8 LIQUID-PROPELLANT ROCKETS

In liquid-fuelled rockets either bipropellants, such as hydrogen-oxygen kerosine-oxygen and hydrazine-fuming nitric acid , or monopropellants may be used. The power or combustion chamber of a liquid propellant rocket consists of the injectors, combustion chamber and nozzle, the general arrangement being that shown in Figure 5.2.

The development of injectors has by necessity followed empirical design procedures and consequently there are a number of types available. Essentially, simple pressure jet atomisers may be used as in the shower head or spray types, or atomisation may be realised by jet impingement. This latter type may involve a single fluid or a twin fluid (fuel-oxidant) arrangement. A number of other designs are also used and the basic types are shown in Figure 3.5. In many injectors the nozzle assembly is a flat plate as in Figure 6.17, the nozzles being formed by drilling directly into the plate. The angles of inclination of the jets play an important role in determining the efficiency of atomisation and mixing as well as the velocities of the droplets. A very high level of turbulence exists within the combustion chamber and the processes of vaporisation, mixing and combustion that occur are not delineated.

The combustion chamber is usually characterised by the 'characteristic chamber length', L^*, which may be defined as the length a straight tube would have if it were of the same volume and throat

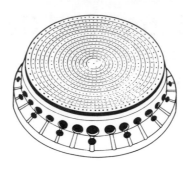

Figure 6.17. *Typical form of liquid propellant injector plate (after Sutton, 1963).*

diameter as the combustor. It can therefore be expressed as

$$L^* = V_c/A_t$$

where V_c is the combustion chamber volume and A_t is the throat area.

Very high combustion intensities are achieved in rocket engines. Thus, for example, the first stage of the Apollo II (and subsequent Apollo rockets) is equipped with five F-1 engines which are powered by the spray combustion of kerosine and liquid oxygen. The flow rates for each motor are 808.5 kg s^{-1} for the fuel and 1875.5 kg s^{-1} for the oxidiser, the total propellant flow being 9.7 x 10^6 kg h^{-1}. The engine power is 8.5 MW and this corresponds to a combustion intensity of about 250 MW m^{-3}.

The subsequent stages are powered by J-2 engines which involve the spray combustion of hydrogen and oxygen. Here the fuel flows for each engine are 37 kg s^{-1} of liquid hydrogen and 192 kg s^{-1} of liquid oxygen. The final stage is powered by a hydrazine derivative-fuming nitric acid propellant combination.

Liquid rocket engines may also exhibit instabilities. These may result from pressure fluctuations in the injector feed line, from enthalpy instability of the type outlined in the previous section, or acoustic instabilities. The injector feed instabilities may readily be overcome by using a carefully engineered injection system.

7 The Nature and Control of Pollutants from Spray Combustion

The pollutants that are formed during the combustion of sprays of liquid fuels are the particulate materials smoke, carbon and unburned droplets, and the gaseous pollutants carbon monoxide, the nitrogen oxides (NO_x), sulphur oxides (SO_x) and unburned hydrocarbon (UHC). The way in which they are formed is indicated in Figure 7.1 and it is clear that the amount of pollutants formed is very much dependent upon the type of application. However, in this chapter the mechanism of pollutant formation and the principles of its control will be considered in a general sense.

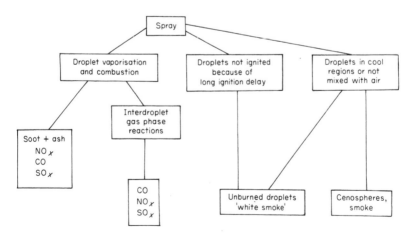

Figure 7.1. Diagrammatic representation of the formation of pollutants during spray combustion.

Whilst it is desirable that the production of combustion-generated pollutants is minimised, it is also essential that whatever method of control is employed the full energy potential of the fuel is still realised, that is, there is no reduction in the overall efficiency of the combustion process or of the utilisation of the energy produced. This aspect has been particularly emphasised recently because of increased fuel costs and because of the need for the conservation of fuel resources. Improvement of the overall efficiency, either directly or indirectly, automatically decreases the yields of pollutants since the total volume of combustion products is reduced.

Emissions of pollutants from spray-fired combustion equipment are subject to legislation in most countries. In the case of stationary equipment in the United Kingdom this legislation is concerned only with the level of the emission of particulates; in the USA, Japan and some European countries the levels of particulates and certain gaseous species are controlled.

In the case of diesel engines a number of countries have legislation controlling the emission of smoke and in 1976, in the USA, legislation will come into force controlling both the emission of smoke and of the pollutants NO_x, UHC and CO. Likewise in 1980 legislation will come into force in the USA restricting the emission from aircraft.

7.1   THE FORMATION AND CONTROL OF PARTICULATE MATERIAL

7.1.1 The General Features of Smoke Formation

The particulate material found in the combustion products from spray systems may be termed carbon, smoke, soot or stack solids depending upon the particular application. In general it consists of three groups of products. Firstly, smoke which is formed via a gas phase combustion/ pyrolysis process. Secondly, it may contain cenospheres which are produced from cracked fuel or carbon together with any ash present in the fuel. These two forms of pollution are commonly termed black smoke. Finally, it may contain any unburned droplets although this is only rarely a problem, being mainly encountered in incorrectly adjusted diesel engines, where it is termed 'white smoke', or sometimes in aviation gas turbines particularly under take-off conditions.

The detailed mechanism of soot formation by means of the gas phase route is complicated and is by no means completely established at the present time. However, the general features are now understood and these are now outlined.

During the combustion of paraffinic hydrocarbons it is generally
accepted that the hydrocarbon radicals that play an important role
during the combustion of hydrocarbons are decomposed (degraded) forming
ethylene and acetylene.   The acetylene thus produced then polymerises
to form polyacetylenes, these polyacetylenes then form soot particles
which coagulate to give the final products.   The nature of the final
product is dependent upon the composition of the fuel oil, particularly
with respect to the concentrations of aromatics and asphaltenes.
However, for a predominantly paraffinic hydrocarbon the mechanism can be
expressed as follows:

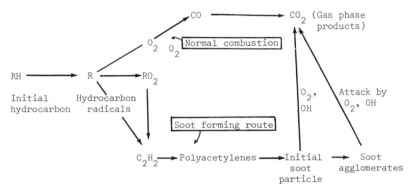

The properties of the soot produced are very much dependent upon the
residence times and the temperatures.   Typically the product contains
90-98% (wt) carbon and only small amounts of oxygen, the particle sizes
vary widely but are usually in the range of 10-1000 nm.   If aromatics
are present the properties may markedly change as well as the yield,
which is enhanced.   The spray combustion of specially selected fuel
oils of high aromatic content is used for the manufacture of carbon
black.   In this case the properties of the final product, particularly
in relation to particle size, can be controlled by the reaction
conditions (Williams, 1973).

For soot to be produced, the fuel-rich conditions must be attained,
otherwise the 'normal combustion' route above is followed and little or
no soot results.   In droplet or spray combustion this condition is
always achieved in part since the region between the droplet surface and
the surrounding flames is always fuel-rich.   Inevitably, soot is always
produced in this region during the combustion of hydrocarbons and the
flame zone always exhibits a yellow luminosity.   The amount of soot
produced in this way may be greatly reduced if the droplet is burning

103

under wake flame conditions which approximate to premixed flames (see Section 4.6) and soot yields may be correlated with the injection velocity and the droplet size. It should be noted that the velocity of the droplet relative to the air at which transition to a wake flame occurs is proportional to $d^{\frac{1}{2}}$ and therefore small droplets are less prone to smoke formation. High injection velocities, whether actual or relative to the incoming air, should also be realised to minimise smoke formation.

Carbon or cenosphere formation from the droplets themselves can also occur but generally this mechanism is only applicable to medium and heavy fuel oils. Lighter fuels such as aviation and diesel fuels generally only produce smoke by means of the gas phase mechanism. In the case of the heavier fuel oils the yield of particulate material is dependent upon the opportunity the droplet has whilst in the combustion chamber to undergo liquid phase cracking. Obviously if the droplets pass through regions which are oxygen-deficient and are at high temperatures, then cenospheres will result. Their mass can be quite substantial, being up to 20% of the mass of the original droplet. This is greatly facilitated as the droplet diameter increases and the concentration of stack solids can readily be shown to increase with droplet size as illustrated in Figure 7.2.

Other important factors are the chemical composition of the oil, particularly in the case of very heavy fuel oils where the asphaltene content may be significant.

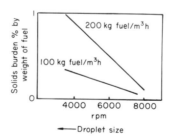

Figure 7.2.   *Variation of the emission of stack solids with droplet size (after Gills, 1972).*

7.1.2   The Formation of Ash

Ash is the inorganic residue left after the combustion of a spray and it originates from materials originally present in the crude oil.

The most common constituents are sodium, vanadium, calcium, magnesium, iron, nickel and silica, the first two being the most important. The sodium is usually present as the chloride, originating from brine associated with the oil; the vanadium is present as soluble organometallic compounds. Consequently, the sodium may be removed to some extent by washing but it is extremely difficult to remove the vanadium.

The major problems resulting from the ash content are concerned with corrosion rather than pollution. The ash deposits on metal surfaces as vanadium or sodium-vanadium compounds reducing heat transfer and causing corrosion. In certain areas in boilers, such as superheaters, the deposits are molten and this retains large ash particles. The melting points of the sodium-vanadium compounds are: α-phase, V_2O_5, m.p. $673^{\circ}C$; β-phase, Na_2O, V_2O_4, $5V_2O_5$, $659^{\circ}C$; γ-phase, $5Na_2O$, V_2O_4, $11V_2O_5$, $577^{\circ}C$ and δ-phase, Na_2O, V_2O_5 (sodium metavanadate), $630^{\circ}C$.

Because ash is concentrated in the residual fuel oils its presence precludes their use in gas turbines on account of their deposition on the turbine blades. It also limits its use in other i.c. engines. On the other hand, since residual fuel oils are largely used in large installations, such as thermal power stations, its presence does not cause any serious pollution problems since it is removed by the usual stack gas cleaning processes by means of electrostatic precipitators and solids arrestors.

7.1.3   Methods of Reducing Emissions of Particulate Materials by Controlling the Combustion Process

Generally the formation of soot or carbon can be minimised by means of the provision of good mixing resulting from turbulence, the maintenance of a high temperature and a sufficient residence time. These three requirements, turbulence, temperature and time are called the 3-t rule. However, the scale of turbulence is also important since smoke and unburned hydrocarbons can originate from turbulent eddies which are fuel-rich.

In the case of stationary combustion plant the levels of excess air are kept very low so as to produce higher flame temperatures and hence greater combustion intensity and efficiency. In addition, the yields of SO_3 (see next section) are also minimised. However, as the excess air is reduced, soot formation results and this limits the reduction possible although low levels of excess air (e.g. 0.3%) may be achieved in well-mixed and well-controlled systems. Good control is essential in such circumstances because any small reduction in the air supply would result in soot formation. However, in much of the installed plant this is not possible and soot formation during low excess air operation may be achieved by means of additives.

Two groups of additives are used to minimise smoke formation. The first of these involves metallic compounds, particularly inorganic compounds of manganese, magnesium or barium although other similar (Groups I or II) metals have some effect. The barium compounds were used experimentally for a number of years with diesel engines, but this application has been terminated because of the toxicity of barium compounds and also because of advances in reducing smoke by improved designs of the combustion chamber. Generally this has involved better mixing by carefully controlled combustion chamber swirl.

In the case of stationary plant and marine applications manganese compounds have increasingly been used. Generally these are added as soluble organo-compounds in a carrier so that the concentration of metal is only one part per several thousand of the oil, thus minimising any pollution problems caused by the additive. The mode of action of the additive is uncertain, but mainly involves the following two functions. These are the reduction of the agglomeration of soot particles because of ionisation effects and the catalytic combustion of any cenospheres. Thus, it influences both the gas phase and liquid phase routes to carbon formation.

The second group of additives are non-metallic and operate by inhibiting the liquid phase polymerisation that causes cenosphere production. Although originally developed ten years ago for diesel engines, they are increasingly being used for boiler plant. Again, in common with the manganese additives, fuel economy may be substantially improved because of the reduction in the excess air requirements. Furthermore, vanadium deposits are reduced.

The formation of carbon may also be minimised by the addition of water to the fuel oil. The presence of steam during the combustion of fuel oils, as is the case when steam is used for atomisation, is beneficial in reducing smoke, since not only does it promote good mixing but it also provides a source of OH radicals which prevent the soot precursors forming. If water is emulsified with the fuel oil this further assists, since the water present in the oil causes disruptive burning and the droplets are effectively further atomised. This technique is extremely promising for stationary plant but because of the necessity of carrying water supplies is not attractive in engine applications.

7.1.4 Method of Reducing Emissions of Particulate Materials by Controlling the Combustion Conditions

The combustion conditions that can be varied are: (a) the degree of swirl; (b) the extent of recirculation; (c) the combustion intensity; (d) the quality of atomisation, and (e) the spatial distribution of the spray in so much as it determines the local fuel-air ratio.

The adjustment of any of these parameters that will make the spray behave as a premixed gaseous flame will result in the reduction of smoke. In this respect, the key factors are the droplet size, the degree of mixing, the relative droplet to gas velocities and the gas composition, since these control whether envelope diffusion flame combustion is replaced to some extent by wake flames (Gills, 1973; Sjorgren, 1973).

The degree of swirl imparted to the combustion air has a considerable influence. Low levels of swirl produce considerable amounts of smoke because of poor mixing and increasing swirl decreases soot formation. Overswirling, however, results in the increased formation of gas phase soot.

Recirculation of combustion gases will, to a certain degree, reduce soot. This can be achieved either by swirl or by actually directing the gases from the flue to the air intake of the burner. In both cases the net result is that hot vitiated gases are mixed with the incoming air increasing its temperature but slightly reducing the oxygen content. The effect of this is to promote droplet vaporisation and to prevent diffusion-controlled droplet burning since extinction or wake-flame combustion occurs.

The other factors, droplet size and spatial disposition, have limitations in what can be achieved with a particular style of atomiser, and changes in basic design, say from pressure jet to air blast atomisers, are necessary to reduce smoke.

7.2 THE FORMATION OF CARBON MONOXIDE

Carbon monoxide is found in the combustion products of all carbonaceous fuels and generally in concentrations somewhat above that expected from equilibrium considerations. For any system which is in equilibrium the carbon monoxide concentration is given by the overall reaction

$$CO_2 \xrightleftharpoons{1} CO + \tfrac{1}{2}O_2 \qquad K_1 \qquad\qquad (i)$$

and thus
$$[CO] = K_1 \frac{[CO_2]}{[O_2]^{\frac{1}{2}}} \qquad (7.1)$$

It is clear that the equilibrium level of carbon monoxide is dependent on the temperature and the level of excess air. Low levels of excess air result in higher concentrations of carbon monoxide if the temperature is maintained constant.

Carbon monoxide is formed in flames by the rapid oxidation of hydrocarbons by oxygen in the reaction zone. The carbon monoxide is subsequently slowly oxidised to carbon dioxide by the reactions (2) and (3) which combined form the water-gas equilibrium reaction:

$$CO + OH \overset{2}{\rightleftharpoons} CO_2 + H \qquad (ii)$$

$$H + H_2O \overset{3}{\rightleftharpoons} H_2 + OH \qquad (iii)$$

$$\overline{CO + H_2O \overset{4}{\rightleftharpoons} CO_2 + H_2} \qquad (iv)$$

Since the carbon monoxide is formed rapidly in the reaction zone but only slowly consumed, the concentrations of carbon monoxide present in the reaction zone are above the equilibrium values. The slow conversion of carbon monoxide to carbon dioxide in the post-flame gases is termed after-burning. The rate of this conversion is given by

$$\frac{-d(CO)}{dt} = 1.8 \times 10^{10}\ (F_{CO})(F_{O_2}^{\frac{1}{2}})(F_{H_2}O^{\frac{1}{2}}) \frac{P}{(RT)^2} \exp(-25\ 000/RT)\ \text{dm mol}^{-1}\text{s}^{-1} \qquad (7.2)$$

where F_x is the mole fraction of x and P the pressure in bar. This expression due to Williams *et al.* (1968) can be integrated to give the concentration of CO at any time.

If the time available for burn-out of the carbon monoxide is short, as in small combustion chambers, the concentrations of carbon monoxide in the burned gases are higher than for large units. Thus the CO levels for small units are about 15 p.p.m. compared with 0.3 p.p.m. for large combustion chambers of the type found in power stations. Low temperatures also promote higher levels of carbon monoxide ; up to 1% may be produced in badly adjusted systems which are running fuel-rich.

In engines the carbon monoxide concentrations are dependent on the residence times (which are generally short particularly in piston engines), on the temperature and on the fuel-air ratio. Generally the highest levels of carbon monoxide are produced under idling conditions when combustion chamber temperatures are low and mixing is poor.

7.3 POLLUTANTS ORIGINATING FROM SULPHUR PRESENT IN LIQUID FUELS

7.3.1 The Formation of Sulphur Dioxide

All petroleum products contain organo-sulphur compounds which are present as sulphides, disulphides or cyclic compounds. Their nature and concentration is dependent upon the origin of the crude oil, but the highest concentrations are found in residual fuels and it is only these fuels that present any problem from the environmental point of view. In fuel oils used in the U.K. the sulphur content generally ranges between 0.1% in kerosene to about 3.0% weight in heavy fuel oils.

On combustion the sulphur compounds are rapidly converted to SO_x in the flame zone as indicated in Figure 7.3. The extent of SO_2 formation (in p.p.m.) is given approximately by 510 x % sulphur in the oil for stoichiometric combustion, so that, typically, SO_2 concentrations in the stack gases are in the range of 200-2000 p.p.m. The sulphur dioxide produced is undesirable for a number of reasons, but the principal one is that it or products derived from it (SO_3, H_2SO_4 aerosol) are air pollutants damaging health and causing corrosion (Williamson, 1973). Since the removal of the SO_2 from the stack gases is a difficult and costly operation, its presence is the principal factor determining chimney heights for stationary installations. Details of methods of estimating chimney heights and the problems associated with dispersion are given in a number of accounts of which Nonhebel (1973) and Williamson (1973) are convenient compilations of the relevant data.

7.3.2 The Formation of Sulphur Trioxide

Sulphur trioxide may be formed from the sulphur dioxide produced initially, the mixture of SO_2 and SO_3 being termed SO_x. Sulphur trioxide is undesirable in that it may react with water to form sulphuric acid. Since the acid dewpoint, or temperature at which condensation occurs, may be as high as 150°C, condensation may occur in economisers and air heaters resulting in the corrosion of metal surfaces and the formation of deposits of sulphates. This problem can be overcome by maintaining all metallic surfaces above the acid dewpoint, by minimising the formation of sulphur trioxide or by the use of additives.

The mechanism of formation of SO_3 is now well established and involves the association reaction with an oxygen atom, namely

$$SO_2 + O = SO_3 \qquad E \sim 0 \text{ kJ mol}^{-1} \qquad (v)$$

or by the reaction

$$SO_2 + O_2 = SO_3 + O \qquad E \sim 300 \text{ kJ mol}^{-1} \qquad (vi)$$

although the reaction may be catalysed by solid surfaces, particularly carbon and iron.

The reactions are moderately rapid under most flame situations and equilibrium is fairly rapidly established so that the concentrations of SO_3 in stack gases may readily be calculated from the equilibrium:

$$SO_2 + \tfrac{1}{2}O_2 \overset{7}{\rightleftharpoons} SO_3 \qquad K_7 \qquad (vii)$$

where $K_7 = \dfrac{\left[SO_3\right]}{\left[SO_2\right]\left[O_2\right]^{\frac{1}{2}}}$ \hfill (7.3)

or $\left[SO_3\right] = K_7\left[SO_2\right]\left[O_2\right]^{\frac{1}{2}}$

Since the concentration of SO_2 can be readily calculated from the sulphur content of the fuel and since the oxygen content in the burned gases is known by direct measurement or by calculation, the amount of SO_3 present can be deduced. It is clear that the concentration of SO_3 increases with the level of excess air (i.e. with $\left[O_2\right]^{\frac{1}{2}}$). Typical values are shown in Figure 7.3 and it is seen that the concentration of SO_3 in flue gases forms about 0.2 to 3% of the total sulphur oxides present and thus SO_3 rarely exceeds 50 p.p.m. Values of K_7 enabling this calculation to be undertaken are given in Appendix 4.

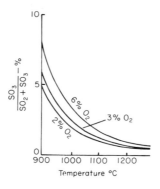

Figure 7.3. *Equilibrium concentrations of sulphur trioxide as a percentage of total sulphur oxides.*

110

7.3.3   The Formation of Acid Smuts

This problem is restricted to stationary combustion units in which carbonaceous residues formed as a result of spray combustion may be carried to the wall of a flue and deposited on the wall at some part where the surface temperature is such that sulphuric acid from the combustion gases has already condensed on it.   Deposits of acid-soaked carbonaceous materials then build up but due to changes in temperature or gas flow small particles break off and are emitted with the gases from the chimney.   These are called 'acid smuts' or 'acid soots' and are a serious pollutant in that they are deposited in the near vicinity of the chimney and they are extremely corrosive and can do considerable damage.

The emission of acid smuts can be minimised by the following methods:

(a) By reducing the formation of sulphuric acid by either using fuels with a low sulphur content or minimising the concentration of SO_3 by reducing the excess air as previously outlined.

(b) By reducing the amount of unburned material by either a change in burner design, or operating conditions as previously described, or by preventing their build-up in the stack.

(c) By the use of additives.   In this technique materials that react with SO_3 or H_2SO_4 are added.   Originally, powdered dolomite $(MgCO_3.CaCO_3)$ was added but this has been almost totally discontinued in favour of the addition of liquid additives or solids in liquid suspension.   The attraction of this technique is that the additive may be injected by a metering pump.

At the present time, a common form of the additive is a slurry of magnesium oxide sometimes combined with aluminium hydroxide.   Its action is two-fold.   It reacts with the SO_3 produced to produce magnesium sulphite and the magnesium also coats any surfaces that are capable of catalysing the conversion of SO_2 to SO_3.   The aluminium modifies the ash produced so that any deposits readily flake off and it also reduces the formation of soot and carbon.

7.4   THE FORMATION AND CONTROL OF OXIDES OF NITROGEN

7.4.1   The Mechanism of Formation of NO_x

During combustion of fuels with air, a small part of the nitrogen present in the air or in the fuel itself reacts with oxygen to form nitric oxide in the flame gases.   This nitric oxide reacts further in the flame or when the combustion products leave the combustion unit to form NO_2 and to a limited extent N_2O_4; the mixture of these oxides of nitrogen so formed is called NO_x.

The formation of NO_x in spray flames involves two major routes. The first is the well established thermal route termed the Zeldovich mechanism. The second involves the reaction of organic or nitrogen compounds present in the oil and its contribution is dependent upon the origins of the oil; this aspect is discussed later.

In the thermal mechanism, oxygen atoms, which are present in the flame zone, and combustion products react thus:

$$O + N_2 \overset{8}{\rightleftharpoons} NO + N \qquad\qquad\qquad (viii)$$

$$N + O_2 \overset{9}{\rightleftharpoons} NO + O \qquad\qquad\qquad (ix)$$

$$N_2 + O_2 \overset{10}{\rightleftharpoons} 2NO \qquad\qquad K_{10} \qquad\qquad (x)$$

hence $\qquad \left[NO\right] = \sqrt{(K_{10}\left[N_2\right]\left[O_2\right]})$ $\qquad\qquad\qquad (7.4)$

Values of K_{10} may be derived from the JANAF tables. The rate of reaction 8 is such that it is rate controlling. Furthermore, it is very temperature dependent, so that nitric oxide is only formed in high temperature gases. Since the overall reaction is slow, equilibrium concentrations of NO are only built up in situations where there is a long residence time, that is in large boilers. In smaller combustion units the nitric oxide concentration is limited by the residence time as indicated in Equation 7.5. The concentration of nitric oxide, if it is only produced by means of this mechanism, can be calculated from the expression:

$$\frac{d\left[NO\right]}{dt} = 2.8 \times 10^{10}\ \exp\left(\frac{-135\ 000}{RT}\right)\left[N_2\right]\left[O_2\right]^{\frac{1}{2}} - 0.13 \times 10^3\ \exp\left(\frac{-91\ 600}{RT}\right)\left[NO\right]^2\left[O_2\right]^{-\frac{1}{2}}$$

$$(7.5)$$

This expression can be integrated to give the concentrations of nitric oxide produced after any time. The units of concentration are $mol\ dm^{-3}$ and the rate is in $mol\ dm^{-3}s^{-1}$. The time required for the equilibrium levels of NO to be established is such that the actual concentrations attained are only one third to one tenth of the equilibrium concentrations. This of course varies depending upon the circumstances, differing markedly between stationary plant applications operating at

atmospheric pressure and engines which operate at high pressures.   In
practical situations the nitric oxide formed is most radically influenced
by operating temperature but it is also influenced by the level of excess
air and the residence times.   In boilers, the NO_x produced is
approximately proportional to the oil firing rate and so in practice the
NO_x in combustion gases from oil-fired equipment ranges from 100 p.p.m.
for small installations to 1000 p.p.m. for larger units.   The upper
figure is determined to some extent by the amount of fuel nitrogen
compounds present in the oil , as described in the next section, and most
power stations, for example, operate with levels in the 200-500 p.p.m.
range.   It should be noted that the U.S. Federal limit for NO_x emission
is 370 p.p.m. for a stoichiometric fuel-air mixture averaged over two
hours.

In the case of gas turbines the concentrations of NO_x are in the
region of 60-100 p.p.m. but, of course, the gases are greatly diluted
with air.   If the concentrations are expressed relative to a
stoichiometric fuel-air mixture the concentrations are in the region of
200-500 p.p.m., these figures depending upon the load.

In the case of diesel engines, the NO_x concentrations are very
dependent upon the load and mode of operation.   Generally they are about
1000 p.p.m. for idle and 4000 p.p.m. for full load.   The formation of
NO_x and smoke from engines has been summarised in a number of textbooks,
for example, Starkman (1969), Cornelius and Agnew (1970) and
compilations by S.A.E.

7.4.2   The Influence of the Properties of the Spray on NO_x Formation

A considerable quantity of the NO_x produced in spray combustion is
produced by the flames surrounding individual droplets.   Since it has
been shown that for single droplets there is a marked dependence of NO
emission upon the droplet diameter, then the overall properties of the
spray are of significance.   In particular it has been shown that finer
fuel sprays produce less nitric oxide than sprays containing considerable
quantities of large droplets.

A further feature of spray combustion is that generally much of the
combustion occurs under fuel-rich conditions.   In these circumstances a
certain amount of nitric oxide may be produced by the so called 'prompt-
NO' route.   Here, carbon containing free radients react with molecular
nitrogen to form nitric oxide by reactions which probably involve the
following:

113

$$CH + N_2 = HCN + N \qquad \text{(xi)}$$

$$N + O_2 = NO + O \qquad \text{(xii)}$$

Evidence for this comes from the fact that hydrogen cyanide is formed during the combustion of sprays under fuel-rich conditions and is also present in the flame surrounding burning single droplets.

Liquid fuels also contain organic nitrogen compounds such as indoles, carbazoles, pyridines and quinolines. Nitric oxide may also be produced directly from these compounds so that in spray combustion nitric oxide is produced by two routes, this direct formation and by the thermal (Zeldovich) route. Not all the fuel-nitrogen is converted to NO; some produces nitrogen. The relative yields of NO and N_2 depend upon the concentration of the fuel-bound nitrogen in the oil. Low concentrations (<0.05% wt) show high conversion levels whilst in higher concentrations (0.5% wt) the conversion to NO may be only of the order of 50%.

Even so, the presence of fuel-nitrogen in heavy fuel oils may increase the concentration of nitric oxide in the stack gases by several hundred parts per million. Since distillate fuels have only low concentrations of fuel-nitrogen the problem is of little significance in relation to engine fuels.

Little is yet known about the detailed chemical mechanism of the conversion of fuel nitrogen to nitric oxide, but it is established that the process is less temperature dependent than the thermal route, and reduction in flame temperature does not markedly reduce nitric oxide formed in this way.

7.4.3   The Control of NO_x Formation

Since oxygen atom concentrations and flame temperatures are low in fuel-rich situations a number of NO_x control methods have attempted to adopt this technique. Generally this involves two-stage combustion, the first stage of combustion being undertaken under rich conditions and in the second stage the additional air is added to complete combustion. In this way the oxygen atom concentration as well as the peak temperature is reduced and thus the rate of NO formation is reduced. However, a major difficulty is that there is a tendency for soot to be formed and considerable care has to be exercised in its application. In flowing types of combustion chambers such as furnaces, boilers and gas turbines this method is readily applied by operating one or more burners rich and adding additional air later on. This is staged combustion.

Alternatively, in multi-burner arrays some burners are run rich and other burners are used to provide the additional air. In reciprocating c.i. engines the use of indirect injection is one means of controlling NO_x production because combustion initially takes place under rich conditions. However, even in direct injection engines, a considerable part of combustion occurs under rich conditions and if carefully controlled the necessary low level of NO_x may be achieved.

An alternative approach is to use external exhaust gas recirculation in which some of the exhaust gases are recycled. This technique is applicable to stationary units and diesel engines but not gas turbines, at least, in propulsion applications. Here, the recirculation of oxygen deficient combustion products results in some reduction in the flame temperatures and there is a consequential reduction in NO formation. In addition, some of the NO present in the recirculated gases is destroyed on its passage through the flame zone by the reverse of reaction 6. Furthermore soot formation may be reduced as previously described. However, the additional equipment necessary for the gas recirculation makes its application unsatisfactory for engine use and the use of swirl and carefully engineered combustion chambers is a much more satisfactory solution.

References

ASTM. Manual on Measurement and Sampling.

Allinson, J.P. (1973) Criteria for Quality of Petroleum Products. Applied Science Publishers, London.

Allison, C.B. and Faeth, G.M. (1972) Combustion and Flame, 19, 213.

Anson, D. and Tindall, D. (1967) J. Inst. Fuel, 40, 551.

Beér, J.M. and Chigier, N.A. (1973) Combustion Aerodynamics. Applied Science Publishers, London.

Crookes, R.J., Janota, A. and Tan, K.J. (1973) Combustion Institute European Symposium, Academic Press, London.

Cornelius, W. and Agnew, W.G. (1970) Emissions from Continuous Systems. Plenum Press, London.

Dickerson, R.A. and Schuman, M.D. (1956) J. of Spacecraft, 2, 99.

Dombrowski, N., Horne, W. and Williams, A. (1974). Combustion Science and Technology, 27, 111.

Dombrowski, N. and Munday, G. (1968) Biochemical and Biological Engineering Science, Academic Press, New York.

Energy for the Future, (1973). Institute of Fuel, London.

Gaydon, A.G. and Wolfhard, H.G. (1970) Flames. Their Structure, Radiation and Temperature. Chapman and Hall, London.

Gills, B.G. (1972) First National Convention, Combustion and Environment, The Institute of Fuel, London.

Godsave, G.A.E. (1953) Fourth Symposium on Combustion. Williams and Wilkins, Baltimore, p.818.

Hobson and Pohl (1974) Modern Petroleum Technology. 4th Edition, Applied Science Publishers, London.

Hottel, H.C. and Howard, J.B. (1971) New Energy Technology. The Massachusetts Institute of Technology, Cambridge, Massachusetts.

Houghton, H.G. (1950)   Section on Spray Nozzles, Chemical Engineers' Handbook, (Ed. J.H. Perry), McGraw Hill, London.

JANAF Thermochemical Tables (1968-) Dow Chemical Company, Midland, Michigan.

Khan, I.M., Greeves, G. and Probert, D.M. (1973)   Fourteenth Symposium on Combustion, The Combustion Institute, Pittsburgh, Pennsylvania.

McCreath, C.G. and Chigier, N.A. (1973)   Fourteenth Symposium on Combustion, The Combustion Institute, Pittsburgh, Pennsylvania, p.1355.

Mizutani, Y. and Ogasawara, M. (1965)   Int. J. Heat and Mass Transfer, 8, 921.

Nonhebel, G. (1973)   Gas Purification Processes for Air Pollution Control. Newnes-Butterworths, London, p.630.

Nuruzzaman, A.S.M., Siddall, R.G. and Beér, J.M. (1971)   Chem. Eng. Sci., 26, 1635.

Probert, R.P. (1946)   Phil. Mag., 37, 94.

Putnam, A.A. and Thomas, R.E. (1957)   Injection and Combustion of Liquid Fuels. WADC Technical Report, 56-344.

Pye, J.W. (1970)   J. Inst. Fuel, 43, 157.

Sjorgren, A. (1973)   Fourteenth Symposium (International) on Combustion, p.919, The Combustion Institute, Pittsburgh, Pennsylvania.

Society of Automotive Engineers.   Vehicle Emissions, Parts I-IV.

Spalding, D.B. (1953)   Fourth Symposium on Combustion.   Williams and Wilkins, Baltimore, p.847.

Spalding, D.B. (1959)   J. Am. Rocket Soc., 29, 828.

Spiers, H.M. (1962)   Technical Data on Fuel.   British National Committee, World Energy Conference.

Standards for Petroleum and its Products (1974)   Institute of Petroleum.

Starkman, E.S. (1969)   Combustion Generated Pollution.   Plenum Press, New York.

Street, P.J. (1974)   Unpublished results.

Sutton, G.P. (1963)   Rocket Propulsion Elements.   3rd Edition, John Wiley, London.

Tanasawa and Tesima (1958)   Bull. of JSME, 1, 36.

Thring, M.W. (1962)   The Science of Flames and Furnaces.   Chapman and Hall, London.

Walton and Prewett (1949)   Proc. Phys. Soc., LX-11, 341.

Weinberg (1953)   Proc. Roy. Soc., A127, 58.

Williams, A. (1973)   Combustion and Flame, 21, 1.

Williams, G.C., Hottel, H.C. and Morgan, A.C. (1968)   Twelfth Symposium
   (International) on Combustion, p.913.   The Combustion Institute,
   Pittsburgh, Pennsylvania.

Williams, F.A. (1965)   Combustion Theory.   Addison-Wesley, Reading,
   Mass.

Williams, S.J. (1973)   Fundamentals of Air Pollution.   Addison-Wesley,
   Reading, Mass.

Wise, H. and Agoston, G.A. (1958)   Literature of the Combustion of
   Petroleum.   American Chemical Society, Washington, D.C., p.125.

Appendix 1 Table of Energy and Power Equivalents

1 metric ton (1000 kg) or tonne of oil = 7.5 barrels oil = 119 x 10^3 kWh

1 kWh = 3412 Btu = 859.845 k cal = 3600 kJ

1 therm = 100 000 Btu = 29.3 kWh = 25 200 k cal = 105 506 kJ

1 Btu h^{-1} = 2.93 x 10^{-4} kW

In addition

1 metric ton oil = 1.7 metric ton coal = 975 m^3 natural gas (approx.)

Redwood I (seconds)	Redwood II (seconds)	Saybolt Universal (seconds)	Saybolt Furol (seconds)	Engler (degrees)	Kinematic (centistokes)
34*	–	37	–	1.25*	3.4
36	–	40	–	1.3*	4.2
38	–	42	–	1.4*	5.0
40	–	45	–	1.45	5.7
45	–	50	–	1.6*	7.5
50	–	57	–	1.8	9.4
60	–	68	–	2.1	12.6
70	–	79	–	2.4	15.5
80	–	92	–	2.7	18.6
90	–	103	–	3.0	21.3
100	–	115	15*	3.4	24.1
120	–	137	17*	4.0	29.2
150	–	171	21*	4.9	36.8
200	–	228	26	6.5	49
250	–	285	31	8.1	62
300	–	342	37	9.8	74
350	36	399	42	11.4	86
400	41	456	48	13.0	99
450	46	513	53	14.7	111
500	51	570	59	16.3	124
550	56	628	65	17.9	136
600	61	684	71	19.5	148
700	71	799	82	22.8	173
800	81	912	94	26.1	198
900	91	1 025	105	29.3	222
1 000	100	1 142	117	32.6	247
1 100	110	1 257	128	35.9	272
1 200	120	1 368	140	39	296
1 400	140	1 599	163	46	346
1 600	160	1 825	186	52	395
1 800	180	2 050	209	59	444
2 000	200	2 280	232	65	493
2 200	220	2 510	255	72	543
2 400	240	2 735	278	78	592
2 600	260	2 965	302	85	642
2 800	280	3 190	325	91	691

Redwood I (seconds)	Redwood II (seconds)	Saybolt Universal (seconds)	Saybolt Furol (seconds)	Engler (degrees)	Kinematic (centistokes)
3 000	300	3 420	348	98	741
3 500	350	3 990	406	114	864
4 000	400	4 560	464	130	987
4 500	450	4 140	522	147	1 112
5 000	500	5 700	580	163	1 235
5 500	550	6 280	639	179	1 359
6 000	600	6 840	696	195	1 482
6 500	650	7 415	754	212	1 605
7 000	700	7 990	814	228	1 730
7 500	750	8 550	869	244	1 850
8 000	800	9 120	928	261	1 975

The above table is sufficiently accurate for normal commercial requirements, but an asterisk (*) indicates that the conversion figure is only approximate in the low viscosity ranges.

Appendix 3 Gamma Functions

The Gamma function may be defined thus

$$\tau(n) = \int_0^\infty e^{-x} x^{n-1} \, dx \qquad \text{(i)}$$

If n is a positive integer, integration by parts show that

$$\tau(n) = (n-1)! \qquad \text{(ii)}$$

For most practical applications this integrand sufficiently approximates the incomplete Gamma functions. These functions are tabulated in standard texts, some values useful to spray combustion systems are listed below.

n	$\tau(n)$
2.0	1.000
2.5	1.329
3.0	2.000
3.5	3.323
4.0	6.000

Other values may be obtained from the relationship $\Gamma(n + 1) = n\Gamma(n)$

Appendix 4 Equilibrium Data for the SO$_2$, SO$_3$ Reaction

$$K_7 = \frac{P_{SO_3}}{P_{SO_2} P_{O_2}^{\frac{1}{2}}}$$

P in bar (1 bar = 1.0135 atmos)

Temperature (K)	Equilibrium constant, K_7
1000	1.932
1100	6.90×10^{-1}
1200	2.97×10^{-1}
1300	1.47×10^{-1}
1400	7.78×10^{-1}
1500	4.86×10^{-2}
1600	3.133×10^{-2}
1700	2.148×10^{-2}
1800	1.545×10^{-2}
1900	1.156×10^{-2}
2000	8.937×10^{-4}

Index

swirl, 81,94

toroidal burner, 40,86,88

transfer number, 55

turn-down ratio, 32

viscosity, 12

volume:
    cumulative, 24
    incremental, 24

water, 13

wax method, 18,19